Tucholsky Wagner Zola Scott Sydow Freud Schlegel
Turgenev Wallace Fonatne Friedrich II. von Preußen
Twain Walther von der Vogelweide Fouqué Freiligrath Frey
Weber Kant Ernst Frommel
Fechner Fichte Weiße Rose von Fallersleben Richthofen
Hölderlin
Engels Fielding Eichendorff Tacitus Dumas
Fehrs Faber Flaubert Eliasberg Ebner Eschenbach
Feuerbach Maximilian I. von Habsburg Fock Eliot Zweig
Ewald Vergil
Goethe Elisabeth von Österreich London
Mendelssohn Balzac Shakespeare Dostojewski Ganghofer
Trackl Lichtenberg Rathenau Doyle Gjellerup
Stevenson Hambruch
Mommsen Tolstoi Lenz Droste-Hülshoff
Thoma von Arnim Hanrieder
Dach Verne Hägele Hauff Humboldt
Reuter Rousseau Hagen Hauptmann Gautier
Karrillon Garschin Defoe Baudelaire
Damaschke Descartes Hebbel
Hegel Kussmaul Herder
Wolfram von Eschenbach Schopenhauer Rilke George
Bronner Darwin Melville Grimm Jerome
Campe Horváth Aristoteles Bebel Proust
Bismarck Vigny Barlach Voltaire Federer Herodot
Gengenbach Heine
Storm Casanova Tersteegen Gilm Grillparzer Georgy
Chamberlain Lessing Langbein Gryphius
Brentano Lafontaine
Strachwitz Claudius Schiller Kralik Iffland Sokrates
Katharina II. von Rußland Bellamy Schilling
Gerstäcker Raabe Gibbon Tschechow
Löns Hesse Hoffmann Gogol Wilde Gleim Vulpius
Luther Heym Hofmannsthal Klee Hölty Morgenstern
Roth Heyse Klopstock Kleist Goedicke
Luxemburg Puschkin Homer Mörike
Machiavelli La Roche Horaz Musil
Navarra Aurel Musset Kierkegaard Kraft Kraus
Lamprecht Kind Kirchhoff Hugo Moltke
Nestroy Marie de France
Laotse Ipsen Liebknecht
Nietzsche Nansen Ringelnatz
Marx Lassalle Gorki Klett Leibniz
von Ossietzky May vom Stein Lawrence Irving
Petalozzi Knigge
Platon Kafka
Sachs Pückler Michelangelo Kock
Poe Liebermann Korolenko
de Sade Praetorius Mistral Zetkin

The publishing house tredition has created the series **TREDITION CLASSICS**. It contains classical literature works from over two thousand years. Most of these titles have been out of print and off the bookstore shelves for decades.

The book series is intended to preserve the cultural legacy and to promote the timeless works of classical literature. As a reader of a **TREDITION CLASSICS** book, the reader supports the mission to save many of the amazing works of world literature from oblivion.

The symbol of **TREDITION CLASSICS** is Johannes Gutenberg (1400 – 1468), the inventor of movable type printing.

With the series, tredition intends to make thousands of international literature classics available in printed format again – worldwide.

All books are available at book retailers worldwide in paperback and in hardcover. For more information please visit: www.tredition.com

tredition was established in 2006 by Sandra Latusseck and Soenke Schulz. Based in Hamburg, Germany, tredition offers publishing solutions to authors and publishing houses, combined with worldwide distribution of printed and digital book content. tredition is uniquely positioned to enable authors and publishing houses to create books on their own terms and without conventional manufacturing risks.

For more information please visit: www.tredition.com

Seasoning of Wood

J. B. (Joseph Bernard) Wagner

Imprint

This book is part of the TREDITION CLASSICS series.

Author: J. B. (Joseph Bernard) Wagner
Cover design: toepferschumann, Berlin (Germany)

Publisher: tredition GmbH, Hamburg (Germany)
ISBN: 978-3-8495-1405-1

www.tredition.com
www.tredition.de

PREFACE [v]

The seasoning and kiln-drying of wood is such an important process in the manufacture of woods that a need for fuller information regarding it, based upon scientific study of the behavior of various species at different mechanical temperatures, and under different drying processes is keenly felt. Everyone connected with the woodworking industry, or its use in manufactured products, is well aware of the difficulties encountered in properly seasoning or removing the moisture content without injury to the timber, and of its susceptibility to atmospheric conditions after it has been thoroughly seasoned. There is perhaps no material or substance that gives up its moisture with more resistance than wood does. It vigorously defies the efforts of human ingenuity to take away from it, without injury or destruction, that with which nature has so generously supplied it.

In the past but little has been known of this matter further than the fact that wood contained moisture which had to be removed before the wood could be made use of for commercial purposes. Within recent years, however, considerable interest has been awakened among wood-users in the operation of kiln-drying. The losses occasioned in air-drying and improper kiln-drying, and the necessity for getting the material dry as quickly as possible after it has come from the saw, in order to prepare it for manufacturing purposes, are bringing about a realization of the importance of a technical knowledge of the subject.

Since this particular subject has never before been represented by any technical work, and appears to have been neglected, it is hoped that the trade will appreciate the endeavor in bringing this book before them, as well as the difficulties encountered in compiling it, as it is the first of [vi] its kind in existence. The author trusts that his efforts will present some information that may be applied with advantage, or serve at least as a matter of consideration or investigation.

In every case the aim has been to give the facts, and wherever a machine or appliance has been illustrated or commented upon, or the name of the maker has been mentioned, it has not been with the intention either of recommending or disparaging his or their work, but has been made use of merely to illustrate the text.

The preparation of the following pages has been a work of pleasure to the author. If they prove beneficial and of service to his fellow-workmen he will have been amply repaid.

THE AUTHOR.

September, 1917

CONTENTS [vii]

LIST OF ILLUSTRATIONS [xi]

FIG.

SEASONING OF WOOD [1]

SECTION I

TIMBER

Characteristics and Properties

Timber was probably one of the earliest, if not the earliest, of materials used by man for constructional purposes. With it he built for himself a shelter from the elements; it provided him with fuel and oft-times food, and the tree cut down and let across a stream formed the first bridge. From it, too, he made his "dug-out" to travel along and across the rivers of the district in which he dwelt; so on down through the ages, for shipbuilding and constructive purposes, timber has continued to our own time to be one of the most largely used of nature's products.

Although wood has been in use so long and so universally, there still exists a remarkable lack of knowledge regarding its nature, not only among ordinary workmen, but among those who might be expected to know its properties. Consequently it is often used in a faulty and wasteful manner. Experience has been almost the only teacher, and theories—sometimes right, sometimes wrong—rather than well substantiated facts, lead the workman.

One reason for this imperfect knowledge lies in the fact that wood is not a homogeneous material, but a complicated structure, and so variable, that one piece will behave very differently from another, although cut from the same tree. Not only does the wood of one species differ from that of another, but the butt cut differs from that of the top log, the heartwood from the sapwood; the wood of quickly-grown sapling of the abandoned field, from [2] that of the slowly-grown, old monarch of the forest. Even the manner in which the tree was cut and kept influences its behavior and quality. It is therefore extremely difficult to study the material for the purpose of establishing general laws.

The experienced woodsman will look for straight-grained, long-fibred woods, with the absence of disturbing resinous and coloring matter, knots, etc., and will quickly distinguish the more porous red or black oaks from the less porous white species, *Quercus alba*. That the inspection should have regard to defects and unhealthy conditions (often indicated by color) goes without saying, and such inspection is usually practised. That knots, even the smallest, are defects, which for some uses condemn the material entirely, need hardly be mentioned. But that "season-checks," even those that have closed by subsequent shrinkage, remain elements of weakness is not so readily appreciated; yet there cannot be any doubt of this, since these, the intimate connections of the wood fibres, when once interrupted are never reestablished.

Careful woods-foremen and manufacturers, therefore, are concerned as to the manner in which their timber is treated after the felling, for, according to the more or less careful seasoning of it, the season checks—not altogether avoidable—are more or less abundant.

There is no country where wood is more lavishly used or criminally neglected than in the United States, and none in which nature has more bountifully provided for all reasonable requirements.

In the absence of proper efforts to secure reproduction, the most valuable kinds are rapidly being decimated, and the necessity of a more rational and careful use of what remains is clearly apparent. By greater care in selection, however, not only will the duration of the supply be extended, but more satisfactory results will accrue from its practice.

There are few more extensive and wide-reaching subjects on which to treat than timber, which in this book refers to dead timber—the timber of commerce—as [3] distinct from the living tree. Such a great number of different kinds of wood are now being brought from various parts of the world, so many new kinds are continually being added, and the subject is more difficult to explain because timber of practically the same character which comes from different localities goes under different names, that if one were always to adhere to the botanical name there would be less confusion, although even botanists differ in some cases as to names. Except in the cases of the older and better known timbers, one rarely takes up two books dealing with timber and finds the botanical names the same; moreover, trees of the same species may produce a much poorer quality of timber when obtained from different localities in the same country, so that botanical knowledge will not always allow us to dispense with other tests.

The structure of wood affords the only reliable means of distinguishing the different kinds. Color, weight, smell, and other appearances, which are often direct or indirect results of structure, may be helpful in this distinction, but cannot be relied upon entirely. Furthermore, structure underlies nearly all the technical properties of this important product, and furnishes an explanation why one piece differs in these properties from another. Structure explains why oak is heavier, stronger, and tougher than pine; why it is harder to saw and plane, and why it is so much more difficult to season without injury. From its less porous structure alone it is evident that a piece of young and thrifty oak is stronger than the porous wood of an old or stunted tree, or that a Georgia or long-leaf pine excels white pine in weight and strength.

Keeping especially in mind the arrangement and direction of the fibres of wood, it is clear at once why knots and "cross-grain" interfere with the strength of timber. It is due to the structural peculiarities that "honeycombing" occurs in rapid seasoning, that checks or cracks extend radially and follow pith rays, that tangent or "bastard" cut stock shrinks and warps more than that which is quarter-sawn. These same peculiarities enable oak to take a better finish than basswood or coarse-grained pine. [4]

Structure of Wood

The softwoods are made up chiefly of tracheids, or vertical cells closed at the ends, and of the relatively short parenchyma cells of the medullary rays which extend radially from the heart of the tree. The course of the tracheids and the rays are at right angles to each other. Although the tracheids have their permeable portions or pits in their walls, liquids cannot pass through them with the greatest ease. The softwoods do not contain "pores" or vessels and are therefore called "non-porous" woods.

The hardwoods are not so simple in structure as softwoods. They contain not only rays, and in many cases tracheids, but also thick-walled cells called fibres and wood parenchyma for the storage of such foods as starches and sugars. The principal structural features of the hardwoods are the pores or vessels. These are long tubes, the segments of which are made up of cells which have lost their end walls and joined end to end, forming continuous "pipe lines" from the roots to the leaves in the tree. Since they possess pores or vessels, the hardwoods are called "porous" woods.

Red oak is an excellent example of a porous wood. In white oak the vessels of the heartwood especially are closed, very generally by ingrowths called tyloses. This probably explains why red oak dries more easily and rapidly than white oak.

The red and black gums are perhaps the simplest of the hardwoods in structure. They are termed "diffuse porous" woods because of the numerous scattered pores they contain. They have only vessels, wood fibres, and a few parenchyma cells. The medullary rays, although present, are scarcely visible in most instances. The vessels are in many cases open, and might be expected to offer relatively little resistance to drying.

Properties of Wood

Certain general properties of wood may be discussed briefly. We know that wood substance has the property of taking in moisture from the air until some balance is [5] reached between the humidity of the air and the moisture in the wood. This moisture which goes into the cell walls hygroscopic moisture, and the property which the wood substance has of taking on hygroscopic moisture is termed hygroscopicity. Usually wood contains not only hygroscopic moisture but also more or less free water in the cell cavities. Especially is this true of sapwood. The free water usually dries out quite rapidly with little or no shrinkage or other physical change.

In certain woods—for example, *Eucalyptus globulus* and possibly some oaks—shrinkage begins almost at once, thus introducing a factor at the very start of the seasoning process which makes these woods very refractory.

The cell walls of some species, including the two already mentioned, such as Western red cedar and redwood, become soft and plastic when hot and moist. If the fibres are hot enough and very wet, they are not strong enough to withstand the resulting force of the atmospheric pressure and the tensile force exerted by the departing free water, and the result is that the cells actually collapse.

In general, however, the hygroscopic moisture necessary to saturate the cell walls is termed the "fibre saturation point." This amount has been found to be from 25 to 30 per cent of the dry wood weight. Unlike *Eucalyptus globulus* and certain oaks, the gums do not begin to shrink until the moisture content has been reduced to about 30 per cent of the dry wood weight. These woods are not subject to collapse, although their fibres become very plastic while hot and moist.

Upon the peculiar properties of each wood depends the difficulty or ease of the seasoning process.

Classes of Trees

The timber of the United States is furnished by three well-defined classes of trees: (1) The needle-leaved, naked-seeded conifers, such as pine, cedar, etc., (2) the broad-leaved trees such as oak poplar, etc., and (3) to an inferior extent by the (one-seed leaf) palms, yuccas, and their allies, which are confined to the most southern parts of the country. [6]

Broad-leaved trees are also known as deciduous trees, although, especially in warm countries, many of them are evergreen, while the needle-leaved trees (conifers) are commonly termed "evergreens," although the larch, bald cypress, and others shed their leaves every fall, and even the names "broad-leaved" and "coniferous," though perhaps the most satisfactory, are not at all exact, for the conifer "ginkgo" has broad leaves and bears no cones.

Among the woodsmen, the woods of broad-leaved trees are known as "hardwoods," though poplar is as soft as pine, and the "coniferous woods" are known as "softwoods," notwithstanding the fact that yew ranks high in hardness even when compared with "hardwoods."

Both in the number of different kinds of trees or species and still more in the importance of their product, the conifers and broad-leaved trees far excel the palms and their relatives.

In the manner of their growth both the conifers and broad-leaved trees behave alike, adding each year a new layer of wood, which covers the old wood in all parts of the stem and limbs. Thus the trunk continues to grow in thickness throughout the life of the tree by additions (annual rings), which in temperate climates are, barring accidents, accurate records of the tree. With the palms and their relatives the stem remains generally of the same diameter, the tree of a hundred years old being as thick as it was at ten years, the growth of these being only at the top. Even where a peripheral increase takes place, as in the yuccas, the wood is not laid on in well-defined layers for the structure remains irregular throughout. Though alike in the manner of their growth, and therefore similar in their general make-up, conifers and broad-leaved trees differ markedly in the details of their structure and the character of their wood.

The wood of all conifers is very simple in its structure, the fibres composing the main part of the wood all being alike and their arrangement regular. The wood of the broad-leaved trees is complex in structure; it is made up of different kinds of cells and fibres and lacks the regularity of arrangement so noticeable in the conifers. This [7] difference is so great that in a study of wood structure it is best to consider the two kinds separately.

In this country the great variety of woods, and especially of useful woods, often makes the mere distinction of the kind or species of tree most difficult. Thus there are at least eight pines of the thirty-five native ones in the market, some of which so closely resemble each other in their minute structure that one can hardly tell them apart, and yet they differ in quality and are often mixed or confounded in the trade. Of the thirty-six oaks, of which probably not less than six or eight are marketed, we can readily recognize by means of their minute anatomy at least two tribes — the white and black oaks. The same is true of the eleven kinds of hickory, the six kinds of ash, etc., etc.

The list of names of all trees indigenous to the United States, as enumerated by the United States Forest Service, is 495 in number, the designation of "tree" being applied to all woody plants which produce naturally in their native habitat one main, erect stem, bearing a definite crown, no matter what size they attain.

Timber is produced only by the Spermatophyta, or seed-bearing plants, which are subdivided into the Gymnosperms (conifers), and Angiosperms (broad-leaved). The conifer or cone-bearing tree, to which belong the pines, larches, and firs, is one of the three natural orders of Gymnosperms. These are generally classed as "softwoods," and are more extensively scattered and more generally used than any other class of timber, and are simple and regular in structure. The so-called "hardwoods" are "Dicotyledons" or broad-leaved trees, a subdivision of the Angiosperms. They are generally of slower growth, and produce harder timber than the conifers, but not necessarily so. Basswood, poplar, sycamore, and some of the gums, though classed with the hardwoods, are not nearly as hard as some of the pines.

CONIFEROUS TREES

WOOD OF THE CONIFEROUS TREES

Examining a smooth cross-section or end face of a well-grown log of Georgia pine, we distinguish an envelope of reddish, scaly bark, a small, whitish pith at the center, and between these the wood in a great number of concentric rings.

Bark and Pith

The bark of a pine stem is thickest and roughest near the base, decreases rapidly in thickness from one to one-half inches at the stump to one-tenth inch near the top of the tree, and forms in general about ten to fifteen per cent of the entire trunk. The pith is quite thick, usually one-eighth to one-fifth inch in southern species, though much less so in white pine, and is very thin, one-fifteenth to one twenty-fifth inch in cypress, cedar, and larch.

In woods with a thick pith, the pith is finest at the stump, grows rapidly thicker toward the top, and becomes thinner again in the crown and limbs, the first one to five rings adjoining it behaving similarly.

What is called the pith was once the seedling tree, and in many of the pines and firs, especially after they have been seasoning for a good while, this is distinctly noticeable in the center of the log, and detaches itself from the surrounding wood.

Sap and Heartwood

Wood is composed of duramen or heartwood, and alburnum or sapwood, and when dry consists approximately of 49 per cent by weight of carbon, 6 per cent of hydrogen, 44 per cent of oxygen, and 1 per cent of ash, which is fairly uniform for all species. The sapwood is the external and [9] youngest portion of the tree, and often constitutes a very considerable proportion of it. It lies next the bark, and after a course of years, sometimes many, as in the case of oaks, sometimes few, as in the case of firs, it becomes hardened and ultimately forms the duramen or heartwood. Sapwood is generally of a white or light color, almost invariably lighter in color than the heartwood, and is very conspicuous in the darker-colored woods, as for instance the yellow sapwood of mahogany and similiar colored woods, and the reddish brown heartwood; or the yellow sapwood of *Lignum-vitae* and the dark green heartwood. Sapwood forms a much larger proportion of some trees than others, but being on the outer circumference it always forms a large proportion of the timber, and even in sound, hard pine will be from 40 per cent to 60 per cent of the tree and in some cases much more. It is really imperfect wood, while the duramen or heartwood is the perfect wood; the heartwood of the mature tree was the sapwood of its earlier years. Young trees when cut down are almost all sapwood, and practically useless as good, sound timber; it is, however, through the sapwood that the life-giving juices which sustain the tree arise from the soil, and if the sapwood be cut through, as is done when "girdling," the tree quickly dies, as it can derive no further nourishment from the soil. Although absolutely necessary to the growing tree, sapwood is often objectionable to the user, as it is the first part to decay. In this sapwood many cells are active, store up starch, and otherwise assist in the life processes of the tree, although only the last or outer layer of cells forms the growing part, and the true life of the tree.

The duramen or heartwood is the inner, darker part of the log. In the heartwood all the cells are lifeless cases, and serve only the mechanical function of keeping the tree from breaking under its own great weight or from being laid low by the winds. The darker color of the heartwood is due to infiltration of chemical substances into the cell walls, but the cavities of the cells in pine are not filled up, as is sometimes believed, nor do their walls grow thicker, nor are the walls any more liquified than in the sapwood. [10]

Sapwood varies in width and in the number of rings which it contains even in different parts of the same tree. The same year's growth which is sapwood in one part of a disk may be heartwood in another. Sapwood is widest in the main part of the stem and often varies within considerable limits and without apparent regularity. Generally, it becomes narrower toward the top and in the limbs, its width varying with the diameter, and being the least in a given disk on the side which has the shortest radius. Sapwood of old and stunted pines is composed of more rings than that of young and thrifty specimens. Thus in a pine two hundred and fifty years old a layer of wood or an annual ring does not change from sapwood to heartwood until seventy or eighty years after it is formed, while in a tree one hundred years old or less it remains sapwood only from thirty to sixty years.

The width of the sapwood varies considerably for different kinds of pine. It is small for long-leaf and white pine and great for loblolly and Norway pines. Occupying the peripheral part of the trunk, the proportion which it forms of the entire mass of the stem is always great. Thus even in old long-leaf pines, the sapwood forms 40 per cent of the merchantable log, while in the loblolly and in all young trees the sapwood forms the bulk of the wood.

The Annual or Yearly Rings

The concentric annual or yearly rings which appear on the end face of a log are cross-sections of so many thin layers of wood. Each such layer forms an envelope around its inner neighbor, and is in turn covered by the adjoining layer without, so that the whole stem is built up of a series of thin, hollow cylinders, or rather cones.

A new layer of wood is formed each season, covering the entire stem, as well as all the living branches. The thickness of this layer or the width of the yearly ring varies greatly in different trees, and also in different parts of the same tree.

In a normally-grown, thrifty pine log the rings are widest near the pith, growing more and more narrow toward [11] the bark. Thus the central twenty rings in a disk of an old long-leaf pine may each be one-eighth to one-sixth inch wide, while the twenty rings next to the bark may average only one-thirtieth inch.

In our forest trees, rings of one-half inch in width occur only near the center in disks of very thrifty trees, of both conifers and hardwoods. One-twelfth inch represents good, thrifty growth, and the minimum width of one two hundred inch is often seen in stunted spruce and pine. The average width of rings in well-grown, old white pine will vary from one-twelfth to one-eighteenth inch, while in the slower growing long-leaf pine it may be one twenty-fifth to one-thirtieth of an inch. The same layer of wood is widest near the stump in very thrifty young trees, especially if grown in the open park; but in old forest trees the same year's growth is wider at the upper part of the tree, being narrowest near the stump, and often also near the very tip of the stem. Generally the rings are widest near the center, growing narrower toward the bark.

In logs from stunted trees the order is often reversed, the interior rings being thin and the outer rings widest. Frequently, too, zones or bands of very narrow rings, representing unfavorable periods of growth, disturb the general regularity.

Few trees, even among pines, furnish a log with truly circular cross-section. Usually it is an oval, and at the stump commonly quite an irregular figure. Moreover, even in very regular or circular disks the pith is rarely in the center, and frequently one radius is conspicuously longer than its opposite, the width of some rings, if not all, being greater on one side than on the other. This is nearly always so in the limbs, the lower radius exceeding the upper. In extreme cases, especially in the limbs, a ring is frequently conspicuous on one side, and almost or entirely lost to view on the other. Where the rings are extremely narrow, the dark portion of the ring is often wanting, the color being quite uniform and light. The greater regularity or irregularity of the annual rings has much to do with the technical qualities of the timber. [12]

Spring- and Summer-Wood

Examining the rings more closely, it is noticed that each ring is made up of an inner, softer, light-colored and an outer, or peripheral, firmer and darker-colored portion. Being formed in the forepart of the season, the inner, light-colored part is termed spring-wood, the outer, darker-portioned being the summer-wood of the ring. Since the latter is very heavy and firm it determines to a very large extent the weight and strength of the wood, and as its darker color influences the shade of color of the entire piece of wood, this color effect becomes a valuable aid in distinguishing heavy and strong from light and soft pine wood.

In most hard pines, like the long-leaf, the dark summer-wood appears as a distinct band, so that the yearly ring is composed of two sharply defined bands—an inner, the spring-wood, and an outer, the summer-wood. But in some cases, even in hard pines, and normally in the woods of white pines, the spring-wood passes gradually into the darker summer-wood, so that a darkly defined line occurs only where the spring-wood of one ring abuts against the summer-wood of its neighbor. It is this clearly defined line which enables the eye to distinguish even the very narrow lines in old pines and spruces.

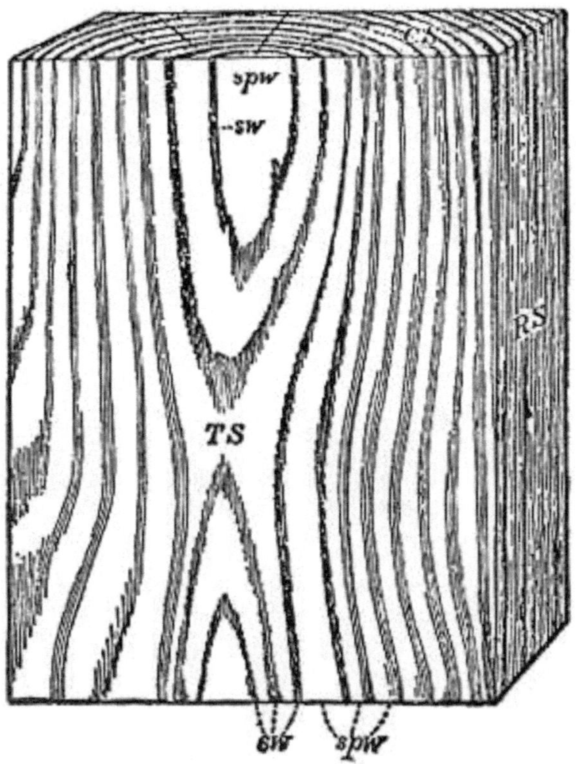

Fig. 1. Board of Pine. CS, cross-section; RS, radial section; TS, tangential section; *sw*, summer-wood; *spw*, spring-wood.

In some cases, especially in the trunks of Southern pines, and normally on the lower side of pine limbs, there occur dark bands of wood in the spring-wood portion of the ring, giving rise to false rings, which mislead in a superficial counting of rings. In the disks cut from limbs these dark bands often occupy the greater part of the ring, and appear as "lunes," or sickle-shaped figures. The wood of these dark bands is similar to that of the true summer-wood. The cells have thick walls, but usually the compressed or flattened form. Normally, the summer-wood forms a greater proportion of the rings in the part of the tree formed during the period of thriftiest growth. In an old tree this proportion is very small in the first two to five rings about the pith, and also in the part next to the bark, the intermediate part showing a greater proportion [13] of summer-wood. It is also greatest in a disk taken from near the stump, and decreases upward in the stem, thus fully accounting for the difference in weight and firmness of the wood of these different parts.

In the long-leaf pine the summer-wood often forms scarcely ten per cent of the wood in the central five rings; forty to fifty per cent of the next one hundred rings, about thirty per cent of the next fifty, and only about twenty per cent in the fifty rings next to the bark. It averages forty-five per cent of the wood of the stump and only twenty-four per cent of that of the top.

Sawing the log into boards, the yearly rings are represented on the board faces of the middle board (radial sections) by narrow parallel strips (see Fig. 1), an inner, lighter stripe and its outer, darker neighbor always corresponding to one annual ring.

On the faces of the boards nearest the slab (tangential or bastard boards) the several years' growth should also appear as parallel, but much broader stripes. This they do if the log is short and very perfect. Usually a variety of pleasing patterns is displayed on the boards, depending on the position of the saw cut and on the regularity of growth of the log (see Fig. 1). Where the cut passes through a prominence (bump or crook) of the log, irregular, concentric circlets and ovals are produced, and on almost all tangent boards arrow or V-shaped forms occur. [14]

Anatomical Structure

Holding a well-smoothed disk or cross-section one-eighth inch thick toward the light, it is readily seen that pine wood is a very porous structure. If viewed with a strong magnifier, the little tubes, especially in the spring-wood of the rings, are easily distinguished, and their arrangement in regular, straight, radial rows is apparent.

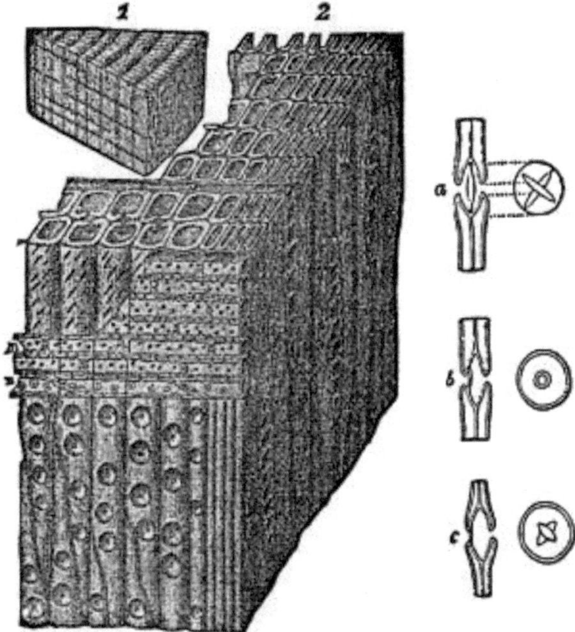

Fig. 2. Wood of Spruce. 1, natural size; 2, small part of one ring magnified 100 times. The vertical tubes are wood fibres, in this case all "tracheids." *m*, medullary or pith ray; *n*, transverse tracheids of ray; *a*, *b*, and *c*, bordered pits of the tracheids, more enlarged.

Scattered through the summer-wood portion of the rings, numerous irregular grayish dots (the resin ducts) disturb the uniformity and regularity of the structure. Magnified one hundred times, a piece of spruce, which is similar to pine, presents a picture like that shown in Fig. 2. Only short pieces of the tubes or cells of which the wood is composed are represented in the picture. The total length of these fibres is from one-twentieth to one-fifth inch, being the smallest near the pith, and is fifty to one hundred times as great as their width (see Fig. 3). They are tapered and closed at their ends, polygonal or rounded and thin-walled, with large cavity, lumen or internal space in the spring-wood, and thick-walled and flattened radially, with the internal space or lumen much reduced in the summer-wood (see right-hand portion of Fig. 2). This flattening, together with the thicker walls of the cells, which reduces the lumen, causes the greater [15] firmness and darker color of the summer-wood. There is more material in the same volume. As shown in the figure, the tubes, cells or "tracheids" are decorated on their walls by circlet-like structures, the "bordered pits," sections of which are seen more magnified as *a*, *b*, and *c*, Fig. 2. These pits are in the nature of pores, covered by very thin membranes, and serve as waterways between the cells or tracheids. The dark lines on the side of the smaller piece (1, Fig. 2) appear when magnified (in 2, Fig. 2) as tiers of eight to ten rows of cells, which run radially (parallel to the rows of tubes or tracheids), and are seen as bands on the radial face and as rows of pores on the tangential face. These bands or tiers of cell rows are the medullary rays or pith rays, and are common to all our lumber woods.

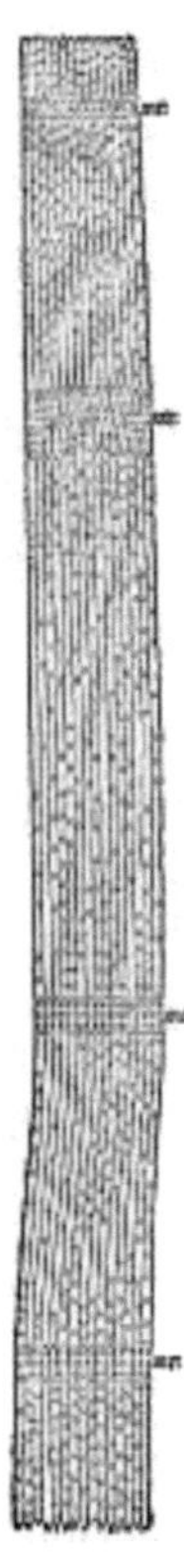

In the pines and other conifers they are quite small, but they can readily be seen even without a magnifier. If a radial surface of split-wood (not smoothed) is examined, the entire radial face will be seen almost covered with these tiny structures, which appear as fine but conspicuous cross-lines. As shown in Fig. 2, the cells of the medullary or pith are smaller and very much shorter than the wood fibre or tracheids, and their long axis is at right angles to that of the fiber.

In pines and spruces the cells of the upper and lower rows of each tier or pith ray have "bordered" pits, like those of the wood fibre or tracheids proper, but the cells of the intermediate rows in the rays of cedars, etc., have only "simple" pits, *i.e.*, pits devoid of the saucer-like "border" or rim. In pine, many of the pith rays are larger than the majority, [16] each containing a whitish line, the horizontal resin duct, which, though much smaller, resembles the vertical ducts on the cross-section. The larger vertical resin ducts are best observed on removal of the bark from a fresh piece of white pine cut in the winter where they appear as conspicuous white lines, extending often for many inches up and down the stem. Neither the horizontal nor the vertical resin ducts are vessels or cells, but are openings between cells, *i.e.*, intercellular spaces, in which the resin accumulates, freely oozing out when the ducts of a fresh piece of sapwood are cut. They are present only in our coniferous woods, and even here they are restricted to pine, spruce, and larch, and are normally absent in fir, cedar, cypress, and yew. Altogether, the structure of coniferous woods is very simple and regular, the bulk being made up of the small fibres called tracheids, the disturbing elements of pith rays and resin ducts being insignificant, and hence the great uniformity and great technical value of coniferous woods. [17]

Fig. 3. Group of Fibres from Pine Wood. Partly schematic. The little circles are "border pits" (see Fig. 2, *a-c*). The transverse rows of square pits indicate the places of contact of these fibres and the cells of the neighboring pith rays. Magnified about 25 times.

LIST OF IMPORTANT CONIFEROUS WOODS

CEDAR

Light soft, stiff, not strong, of fine texture. Sap- and heartwood distinct, the former lighter, the latter a dull grayish brown or red. The wood seasons rapidly, shrinks and checks but little, and is very durable in contact with the soil. Used like soft pine, but owing to its great durability preferred for shingles, etc. Cedars usually occur scattered, but they form in certain localities forests of considerable extent.

(*a*) White Cedars

1. White Cedar (*Thuya occidentalis*) (Arborvitæ, Tree of Life). Heartwood light yellowish brown, sapwood nearly white. Wood light, soft, not strong, of fine texture, very durable in contact with the soil, very fragrant. Scattered along streams and lakes, frequently covering extensive swamps; rarely large enough for lumber, but commonly used for fence posts, rails, railway ties, and shingles. This species has been extensively cultivated as an ornamental tree for at least a century. Maine to Minnesota and northward.

2. Canoe Cedar (*Thuya gigantea*) (Red Cedar of the West). In Oregon and Washington a very large tree, covering extensive swamps; in the mountains much smaller, skirting the water courses. An important lumber tree. The wood takes a fine polish; suitable for interior finishing, as there is much variety of shading in the color. Washington to northern California and eastward to Montana.

3. White Cedar (*Chamæcyparis thyoides*). Medium-sized tree. Heartwood light brown with rose tinge, sapwood [18] paler. Wood light, soft, not strong, close-grained, easily worked, very durable in contact with the soil and very fragrant. Used in boatbuilding cooperage, interior finish, fence posts, railway ties, etc. Along the coast from Maine to Mississippi.

4. White Cedar (*Chamæcyparis Lawsoniana*) (Port Orford Cedar, Oregon Cedar, Lawson's Cypress, Ginger Pine). A very large tree. A fine, close-grained, yellowish-white, durable timber, elastic, easily worked, free of knots, and fragrant. Extensively cut for lumber;

heavier and stronger than any of the preceding. Along the coast line of Oregon.

5. White Cedar (*Libocedrus decurrens*) (Incense Cedar). A large tree, abundantly scattered among pine and fir. Wood fine-grained. Cascades and Sierra Nevada Mountains of Oregon and California.

6. Yellow Cedar (*Cupressus nootkatensis*) (Alaska Cedar, Alaska Cypress). A very large tree, much used for panelling and furniture. A fine, close-grained, yellowish white, durable timber, easily worked. Along the coast line of Oregon north.

(b) Red Cedars

7. Red Cedar (*Juniperus Virginiana*) (Savin Juniper, Juniper, Red Juniper, Juniper Bush, Pencil Cedar). Heartwood dull red color, thin sapwood nearly white. Close even grain, compact structure. Wood light, soft, weak, brittle, easily worked, durable in contact with the soil, and fragrant. Used for ties, posts, interior finish, pencil cases, cigar boxes, silos, tanks, and especially for lead pencils, for which purpose alone several million feet are cut each year. A small to medium-sized tree scattered through the forests, or in the West sparsely covering extensive areas (cedar brakes). The red cedar is the most widely distributed conifer of the United States, occurring from the Atlantic to the Pacific, and from Florida to Minnesota. [19] Attains a suitable size for lumber only in the Southern, and more especially the Gulf States.

8. Red Cedar (*Juniperus communis*) (Ground Cedar). Small-sized tree, its maximum height being about 25 feet. It is found widely distributed throughout the Northern hemisphere. Wood in its quality similar to the preceding. The fruit of this species is gathered in large quantities and used in the manufacture of gin; whose peculiar flavor and medicinal properties are due to the oil of Juniper berries, which is secured by adding the crushed fruit to undistilled grain spirit, or by allowing the vapor to pass over it before condensation. Used locally for construction purposes, fence posts, etc. Ranges from Greenland to Alaska, in the East, southward to Pennsylvania and northern Nebraska; in the Rocky Mountains to Texas, Mexico, and Arizona.

9. Redwood (*Sequoia sempervirens*) (Sequoia, California Redwood, Coast Redwood). Wood in its quality and uses like white cedar. Thick, red heartwood, changing to reddish brown when seasoned. Thin sapwood, nearly white, coarse, straight grain, compact structure. Light, not strong, soft, very durable in contact with the soil, not resinous, easily worked, does not burn easily, receives high polish. Used for timber, shingles, flumes, fence posts, coffins, railway ties, water pipes, interior decorations, and cabinetmaking. A very large tree, limited to the coast ranges of California, and forming considerable forests, which are rapidly being converted into lumber.

CYPRESS

10. Cypress (*Taxodium distinchum*) (Bald Cypress, Black, White, and Red Cypress, Pecky Cypress). Wood in its appearance, quality, and uses similar to white cedar. "Black" and "White Cypress" are heavy and light forms of the same species. Heartwood brownish; sapwood nearly white. Wood close, [20] straight-grain, frequently full of small holes caused by disease known as "pecky cypress." Greasy appearance and feeling. Wood light, soft, not strong, durable in contact with the soil, takes a fine polish. Green wood often very heavy. Used for carpentry, building construction, shingles, cooperage, railway ties, silos, tanks, vehicles, and washing machines. The cypress is a large, deciduous tree, inhabiting swampy lands, and along rivers and coasts of the Southern parts of the United States. Grows to a height of 150 feet and 12 feet in diameter.

FIR

This name is frequently applied to wood and to trees which are not fir; most commonly to spruce, but also, especially in English markets, to pine. It resembles spruce, but is easily distinguished from it, as well as from pine and larch, by the absence of resin ducts. Quality, uses, and habits similar to spruce.

11. Balsam Fir (*Abies balsamea*) (Balsam, Fir Tree, Balm of Gilead Fir). Heartwood white to brownish; sapwood lighter color; coarse-grained, compact structure, satiny. Wood light, not durable or strong, resinous, easily split. Used for boxes, crates, doors, millwork, cheap lumber, paper pulp. Inferior to white pine or spruce, yet often mixed and sold with these species in the lumber market. A medium-sized tree scattered throughout the northern pineries, and cut in lumber operations whenever of sufficient size. Minnesota to Maine and northward.

12. White Fir (*Abies grandis* and *Abies concolor*). Medium- to very large-sized tree, forming an important part of most of the Western mountain forests, and furnishes much of the lumber of the respective regions. The former occurs from Vancouver to California, and the latter from Oregon to Arizona and eastward to Colorado and Mexico. The wood is soft and light, coarse-grained, not unlike the "Swiss pine" of Europe, but [21] darker and firmer, and is not suitable for any purpose requiring strength. It is used for boxes, barrels, and to a small extent for wood pulp.

13. White Fir (*Abies amabalis*). Good-sized tree, often forming extensive mountain forests. Wood similar in quality and uses to *Abies grandis*. Cascade Mountains of Washington and Oregon.

14. Red Fir (*Abies nobilis*) (Noble Fir) (not to be confounded with Douglas spruce. See No. 40). Large to very large-sized tree, forming extensive forests on the slope of the mountains between 3,000 and 4,000 feet elevation. Cascade Mountains of Oregon.

15. Red Fir (*Abies magnifica*). Very large-sized tree, forming forests about the base of Mount Shasta. Sierra Nevada Mountains of California, from Mount Shasta southward.

HEMLOCK

Light to medium weight, soft, stiff, but brittle, commonly cross-grained, rough and splintery. Sapwood and heartwood not well defined. The wood of a light reddish-gray color, free from resin ducts, moderately durable, shrinks and warps considerably in drying, wears rough, retains nails firmly. Used principally for dimension stuff and timbers. Hemlocks are medium- to large-sized trees, commonly scattered among broad-leaved trees and conifers, but often forming forests of almost pure growth.

16. Hemlock (*Tsuga canadensis*) (Hemlock Spruce, Peruche). Medium-sized tree, furnishes almost all the hemlock of the Eastern market. Maine to Wisconsin, also following the Alleghanies southward to Georgia and Alabama.

17. Hemlock (*Tsuga mertensiana*). Large-sized tree, wood claimed to be heavier and harder than the Eastern species and of superior quality. Used for pulp wood, floors, panels, and newels. It is not [22] suitable for heavy construction, especially where exposed to the weather, it is straight in grain and will take a good polish. Not adapted for use partly in and partly out of the ground; in fresh water as piles will last about ten years, but as it is softer than fir it is less able to stand driving successfully. Washington to California and eastward to Montana.

LARCH or TAMARACK

Wood like the best of hard pine both in appearance, quality, and uses, and owing to its great durability somewhat preferred in shipbuilding, for telegraph poles, and railway ties. In its structure it resembles spruce. The larches are deciduous trees, occasionally covering considerable areas, but usually scattered among other conifers.

18. Tamarack (*Larix laricina* var. *Americana*) (Larch, Black Larch, American Larch, Hacmatac). Heartwood light brown in color, sapwood nearly white, coarse conspicuous grain, compact structure, annual rings pronounced. Wood heavy, hard, very strong, durable in contact with the soil. Used for railway ties, fence posts, sills, ship timbers, telegraph poles, flagstaffs. Medium-sized tree, often covering swamps, in which case it is smaller and of poor quality. Maine to Minnesota, and southward to Pennsylvania.

19. Tamarack (*Larix occidentalis*) (Western Larch, Larch). Large-sized trees, scattered, locally abundant. Is little inferior to oak in strength and durability. Heartwood of a light brown color with lighter sapwood, has a fine, slightly satiny grain, and is fairly free from knots; the annual rings are distant. Used for railway ties and shipbuilding. Washington and Oregon to Montana.

PINE

Very variable, very light and soft in "soft" pine, such as white pine; of medium weight to heavy and quite hard in "hard" pine, of which the long-leaf or Georgia [23] pine is the extreme form. Usually it is stiff, quite strong, of even texture, and more or less resinous. The sapwood is yellowish white; the heartwood orange brown. Pine shrinks moderately, seasons rapidly and without much injury; it works easily, is never too hard to nail (unlike oak or hickory); it is mostly quite durable when in contact with the soil, and if well seasoned is not subject to the attacks of boring insects. The heavier the wood, the darker, stronger, and harder it is, and the more it shrinks and checks when seasoning. Pine is used more extensively than any other wood. It is the principal wood in carpentry, as well as in all heavy construction, bridges, trestles, etc. It is also used in almost every other wood industry; for spars, masts, planks, and timbers in shipbuilding, in car and wagon construction, in cooperage and woodenware; for crates and boxes, in furniture work, for toys and patterns, water pipes, excelsior, etc. Pines are usually large-sized trees with few branches, the straight, cylindrical, useful stem forming by far the greatest part of the tree. They occur gregariously, forming vast forests, a fact which greatly facilitates their exploitation. Of the many special terms applied to pine as lumber, denoting sometimes differences in quality, the following deserve attention: "White pine," "pumpkin pine," "soft pine," in the Eastern markets refer to the wood of the white pine (*Pinus strobus*), and on the Pacific Coast to that of the sugar pine (*Pinus lambertiana*). "Yellow pine" is applied in the trade to all the Southern lumber pines; in the Northwest it is also applied to the pitch pine (*Pinus regida*); in the West it refers mostly to the bull pine (*Pinus ponderosa*). "Yellow long-leaf pine" (Georgia pine), chiefly used in advertisements, refers to the long-leaf Pine (*Pinus palustris*).

(*a*) Soft Pines

20. White Pine (*Pinus strobus*) (Soft Pine, Pumpkin Pine, Weymouth Pine, Yellow Deal). Large to very large-sized tree, reaching a height of 80 to 100 feet or more, and in some instances 7 or 8 feet in diameter. For the last fifty years the most important timber tree [24] of the United States, furnishing the best quality of soft pine. Heartwood cream white; sapwood nearly white. Close straight grain, compact structure; comparatively free from knots and resin. Soft, uniform; seasons well; easy to work; nails without splitting; fairly durable in contact with the soil; and shrinks less than other species of pine. Paints well. Used for carpentry, construction, building, spars, masts, matches, boxes, etc., etc., etc.

21. Sugar Pine (*Pinus lambertiana*) (White Pine, Pumpkin Pine, Soft Pine). A very large tree, forming extensive forests in the Rocky Mountains and furnishing most of the timber of the western United States. It is confined to Oregon and California, and grows at from 1,500 to 8,000 feet above sea level. Has an average height of 150 to 175 feet and a diameter of 4 to 5 feet, with a maximum height of 235 feet and 12 feet in diameter. The wood is soft, durable, straight-grained, easily worked, very resinous, and has a satiny luster which makes it appreciated for interior work. It is extensively used for doors, blinds, sashes, and interior finish, also for druggists' drawers, owing to its freedom from odor, for oars, mouldings, shipbuilding, cooperage, shingles, and fruit boxes. Oregon and California.

22. White Pine (*Pinus monticolo*). A large tree, at home in Montana, Idaho, and the Pacific States. Most common and locally used in northern Idaho.

23. White Pine (*Pinus flexilis*). A small-sized tree, forming mountain forests of considerable extent and locally used. Eastern Rocky Mountain slopes, Montana to New Mexico.

(*b*) Hard Pines

24. Long-Leaf Pine (*Pinus palustris*) (Georgia Pine, Southern Pine, Yellow Pine, Southern Hard Pine, Long-straw Pine, etc.). Large-sized tree. This species furnishes the hardest and most durable as [25] well as one of the strongest pine timbers in the market. Heartwood orange, sapwood lighter color, the annual rings are strongly marked, and it is full of resinous matter, making it very durable, but difficult to work. It is hard, dense, and strong, fairly free from knots, straight-grained, and one of the best timbers for heavy engineering work where great strength, long span, and durability are required. Used for heavy construction, shipbuilding, cars, docks, beams, ties, flooring, and interior decoration. Coast region from North Carolina to Texas.

25. Bull Pine (*Pinus ponderosa*) (Yellow Pine, Western Yellow Pine, Western Pine, Western White Pine, California White Pine).

Medium- to very large-sized tree, forming extensive forests in the Pacific and Rocky Mountain regions. Heartwood reddish brown, sapwood yellowish white, and there is often a good deal of it. The resinous smell of the wood is very remarkable. It is extensively used for beams, flooring, ceilings, and building work generally.

26. Bull Pine (*Pinus Jeffreyi*) (Black Pine). Large-sized tree, wood resembles *Pinus ponderosa* and replacing same at high altitudes. Used locally in California.

27. Loblolly Pine (*Pinus tæda*) (Slash Pine, Old Field Pine, Rosemary Pine, Sap Pine, Short-straw Pine). A large-sized tree, forms extensive forests. Wider-ringed, coarser, lighter, softer, with more sapwood than the long-leaf pine, but the two are often confounded in the market. The more Northern tree produces lumber which is weak, brittle, coarse-grained, and not durable, the Southern tree produces a better quality wood. Both are very resinous. This is the common lumber pine from Virginia to South Carolina, and is found extensively in Arkansas and Texas. Southern States, Virginia to Texas and Arkansas.

28. Norway Pine (*Pinus resinosa*) (American Red Pine, Canadian Pine). Large-sized tree, never forming [26] forests, usually scattered or in small groves, together with white pine. Largely sapwood and hence not durable. Heartwood reddish white, with fine, clear grain, fairly tough and elastic, not liable to warp and split. Used for building construction, bridges, piles, masts, and spars. Minnesota to Michigan; also in New England to Pennsylvania.

29. Short-Leaf Pine (*Pinus echinata*) (Slash Pine, Spruce Pine, Carolina Pine, Yellow Pine, Old Field Pine, Hard Pine). A medium- to large-sized tree, resembling loblolly pine, often approaches in its wood the Norway pine. Heartwood orange, sapwood lighter; compact structure, apt to be variable in appearance in cross-section. Wood usually hard, tough, strong, durable, resinous. A valuable timber tree, sometimes worked for turpentine. Used for heavy construction, shipbuilding, cars, docks, beams, ties, flooring, and house trim. *Pinus echinata, palustris*, and *tæda* are very similar in character, of thin wood and very difficult to distinguish one from another. As a rule, however, *palustris* (Long-leaf Pine) has the smallest and most uniform growth rings, and *Pinus tæda* (Loblolly Pine) has the largest. All are apt to be bunched together in the lumber market as Southern Hard Pine. All are used for the same purposes. Short-leaf is the common lumber pine of Missouri and Arkansas. North Carolina to Texas and Missouri.

30. Cuban Pine (*Pinus cubensis*) (Slash Pine, Swamp Pine, Bastard Pine, Meadow Pine). Resembles long-leaf pine, but commonly has a wider sapwood and coarser grain. Does not enter the markets to any extent. Along the coast from South Carolina to Louisiana.

31. Pitch Pine (*Pinus rigida*) (Torch Pine). A small to medium-sized tree. Heartwood light brown or red, sapwood yellowish white. Wood light, soft, not strong, coarse-grained, durable, very resinous. Used locally for lumber, fuel, and charcoal. Coast regions [27] from New York to Georgia, and along the mountains to Kentucky.

32. Black Pine (*Pinus murryana*) (Lodge-pole Pine, Tamarack). Small-sized tree. Rocky Mountains and Pacific regions.

33. Jersey Pine (*Pinus inops* var. *Virginiana*) (Scrub Pine). Small-sized tree. Along the coast from New York to Georgia and along the mountains to Kentucky.

34. Gray Pine (*Pinus divaricata* var. *banksiana*) (Scrub Pine, Jack Pine). Medium- to large-sized tree. Heartwood pale brown, rarely yellow; sapwood nearly white. Wood light, soft, not strong, close-grained. Used for fuel, railway ties, and fence posts. In days gone by the Indians preferred this species for frames of canoes. Maine, Vermont, and Michigan to Minnesota.

REDWOOD (See Cedar)

SPRUCE

Resembles soft pine, is light, very soft, stiff, moderately strong, less resinous than pine; has no distinct heartwood, and is of whitish color. Used like soft pine, but also employed as resonance wood in musical instruments and preferred for paper pulp. Spruces, like pines, form extensive forests. They are more frugal, thrive on thinner soils, and bear more shade, but usually require a more humid climate. "Black" and "White" spruce as applied by lumbermen usually refer to narrow and wide-ringed forms of black spruce (*Picea nigra*).

35. Black Spruce (*Picea nigra* var. *mariana*). Medium-sized tree, forms extensive forests in northwestern United States and in British America; occurs scattered or in groves, especially in low lands throughout the northern pineries. Important lumber tree in eastern United States. Heartwood pale, often with reddish tinge; sapwood pure white. Wood light, [28] soft, not strong. Chiefly used for manufacture of paper pulp, and great quantities of this as well as *Picea alba* are used for this purpose. Used also for sounding boards for pianos, violins, etc. Maine to Minnesota, British America, and in the Alleghanies to North Carolina.

36. White Spruce (*Picea canadensis* var. *alba*). Medium- to large-sized tree. Heartwood light yellow; sapwood nearly white. Generally associated with the preceding. Most abundant along streams and lakes, grows largest in Montana and forms the most important tree of the sub-arctic forest of British America. Used largely for floors, joists, doors, sashes, mouldings, and panel work, rapidly superceding *Pinus strobus* for building purposes. It is very similar to Norway pine, excels it in toughness, is rather less durable and dense, and more liable to warp in seasoning. Northern United States from Maine to Minnesota, also from Montana to Pacific, British America.

37. White Spruce (*Picea engelmanni*). Medium- to large-sized tree, forming extensive forests at elevations from 5,000 to 10,000 feet above sea level; resembles the preceding, but occupies a different station. A very important timber tree in the central and southern parts of the Rocky Mountains. Rocky Mountains from Mexico to Montana.

38. Tide-Land Spruce (*Picea sitchensis*) (Sitka Spruce). A large-sized tree, forming an extensive coast-belt forest. Used extensively for all classes of cooperage and woodenware on the Pacific Coast. Along the sea-coast from Alaska to central California.

39. Red Spruce (*Picea rubens*). Medium-sized tree, generally associated with *Picea nigra* and occurs scattered throughout the northern pineries. Heartwood reddish; sapwood lighter color, straight-grained, compact structure. Wood light, soft, not strong, elastic, resonant, not durable when exposed. Used for flooring, carpentry, shipbuilding, piles, posts, railway [29] ties, paddles, oars, sounding boards, paper pulp, and musical instruments. Montana to Pacific, British America.

BASTARD SPRUCE

Spruce or fir in name, but resembling hard pine or larch in appearance, quality and uses of its wood.

40. Douglas Spruce (*Pseudotsuga douglasii*) (Yellow Fir, Red Fir, Oregon Pine). One of the most important trees of the western United States; grows very large in the Pacific States, to fair size in all parts of the mountains, in Colorado up to about 10,000 feet above sea level; forms extensive forests, often of pure growth, it is really neither a pine nor a fir. Wood very variable, usually coarse-grained and heavy, with very pronounced summer-wood. Hard and strong ("red" fir), but often fine-grained and light ("yellow" fir). It is the chief tree of Washington and Oregon, and most abundant and most valuable in British Columbia, where it attains its greatest size. From the plains to the Pacific Ocean, and from Mexico to British Columbia.

41. Red Fir (*Pseudotsuga taxifolia*) (Oregon Pine, Puget Sound Pine, Yellow Fir, Douglas Spruce, Red Pine). Heartwood light red or yellow in color, sapwood narrow, nearly white, comparatively free from resins, variable annual rings. Wood usually hard, strong, difficult to work, durable, splinters easily. Used for heavy construction, dimension timber, railway ties, doors, blinds, interior finish, piles, etc. One of the most important of Western trees. From the plains to the Pacific Ocean, and from Mexico to British America.

TAMARACK (See Larch)

YEW

Wood heavy, hard, extremely stiff and strong, of fine texture with a pale yellow sapwood, and an orange-red heartwood; seasons well and is quite durable. Extensively [30] used for archery bows, turner's ware, etc. The yews form no forests, but occur scattered with other conifers.

42. Yew (*Taxus brevifolia*). A small to medium-sized tree of the Pacific region.

SECTION III [31]

BROAD-LEAVED TREES

WOOD OF BROAD-LEAVED TREES

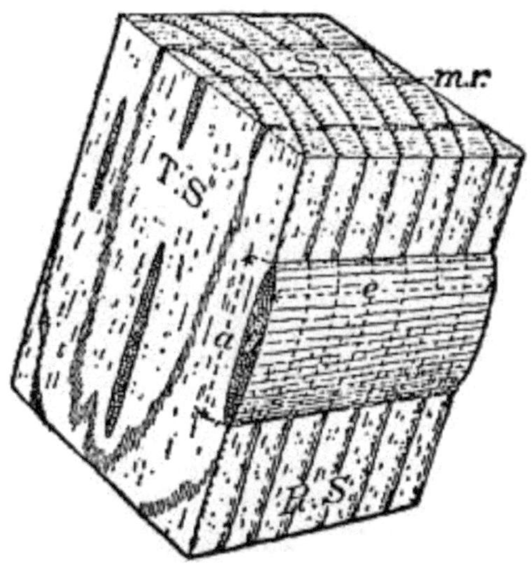

Fig. 4. Block of Oak. CS, cross-section; RS, radial section; TS, tangential section; *mr*, medullary or pith ray; *a*, height; *b*, width; and *e*, length of pith ray.

On a cross-section of oak, the same arrangement of pith and bark, of sapwood and heartwood, and the same disposition of the wood in well-defined concentric or annual rings occur, but the rings are marked by lines or rows of conspicuous pores or openings, which occupy the greater part of the spring-wood for each ring (see Fig. 4, also 6), and are, in fact the hollows of vessels through which the cut has been made. On the radial section or quarter-sawn board the several layers appear as so many stripes (see Fig. 5); on the tangential section or "bastard" face patterns similar to those mentioned for pine wood are observed. But while the patterns in hard pine are marked by the darker summer-wood, and are composed of plain, alternating stripes of darker and lighter wood, the figures in oak (and other broad-leaved woods) are due chiefly to the vessels, those of the spring-wood in oak being the most conspicuous (see Fig. 5). So that in an oak table, the darker, shaded parts are the spring-wood, the lighter unicolored parts the summer-wood. On closer examination of the smooth cross-section of oak, the spring-wood part of the ring is found to be formed in great part of pores; large, round, or oval openings made by the cut through long vessels. These are separated by a grayish [32] and quite porous tissue (see Fig. 6, A), which continues here and there in the form of radial, often branched, patches (not the pith rays) into and through the summer-wood to the spring-wood of the next ring. The large vessels of the spring-wood, occupying six to ten per cent of the volume of a log in very good oak, and twenty-five per cent or more in inferior and narrow-ringed timber, are a very important feature, since it is evident that the greater their share in the volume, the lighter and weaker the wood. They are smallest near the pith, and grow wider outward. They are wider in the stem than limb, and seem to be of indefinite length, forming open channels, in some cases probably as long as the tree itself. Scattered through the radiating gray patches of porous wood are vessels similar to those of the [33] spring-wood, but decidedly smaller. These vessels are usually fewer and larger near the outer portions of the ring. Their number and size can be utilized to distinguish the oaks classed as white oaks from those classed as black and red oaks. They are fewer and larger in red oaks, smaller but much more numerous in white oaks. The summer-wood, except for these radial, grayish patches, is dark colored and firm. This firm portion, divided into bodies or strands by these patches of porous wood, and also by fine, wavy, concentric lines of short, thin-walled cells (see Fig. 6, A), consists of thin-walled fibres (see Fig. 7, B), and is the chief element of strength in oak wood. In good white oak it forms one-half or more of the wood, if it cuts like horn, and the cut surface is shiny, and of a deep chocolate brown color. In very narrow-ringed wood and in inferior red oak it is usually much reduced in quantity as well as quality. The pith rays of the oak, unlike those of the coniferous woods, [34] are at least in part very large and conspicuous. (See Fig. 4; their height indicated by the letter *a*, and their width by the letter *b*.) The large medullary

rays of oak are often twenty and more cells wide, and several hundred cell rows in height, which amount commonly to one or more
inches. These large rays are conspicuous on all sections. They appear as long, sharp, grayish lines on the cross-sections; as short,
thick lines, tapering at each end, on the tangential or "bastard" face,
and as broad, shiny bands, "the mirrors," on the radial section. In
addition to these coarse rays, there is also a large number of small
pith rays, which can be seen only when magnified. On the whole,
the pith rays form a much larger part of the wood than might be
supposed. In specimens of good white oak it has been found that
they form about sixteen to twenty-five per cent of the wood.

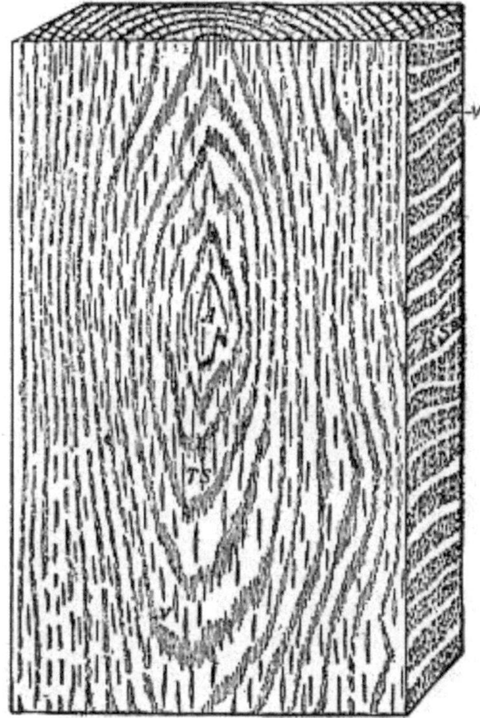

Fig. 5. Board of Oak. CS, cross-section; RS, radial section; TS, tangential section; *v*, vessels or pores, cut through.; A, slight curve in
log which appears in section as an islet.

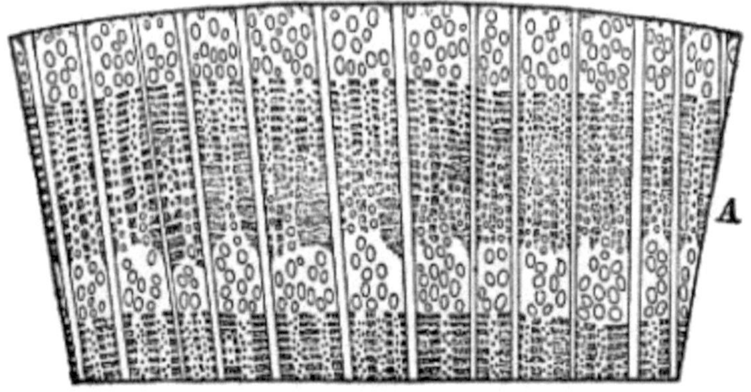

Fig. 6. Cross-section of Oak (Magnified about 5 times).

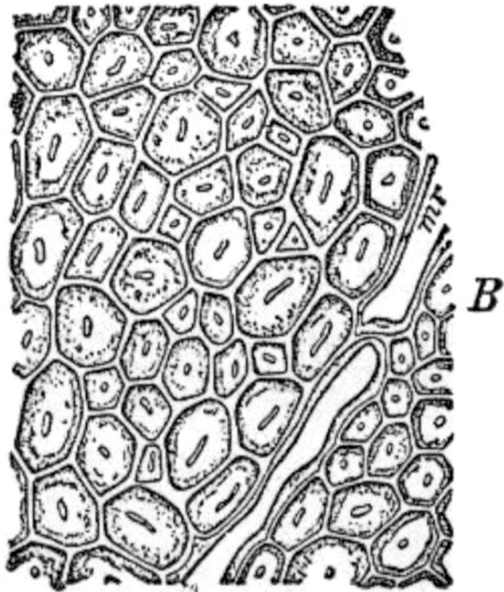

Fig. 7. Portion of the Firm Bodies of Fibres with Two Cells of a Small Pith Ray *mr* (Highly Magnified).

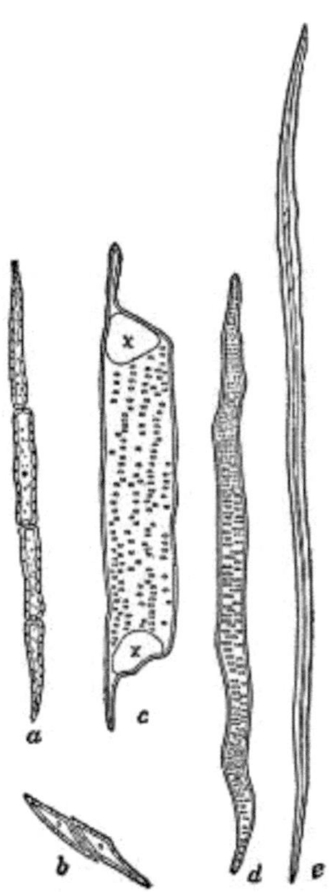

Fig. 8. Isolated Fibres and Cells, *a*, four cells of wood, parenchyma; *b*, two cells from a pith ray; *c*, a single joint or cell of a vessel, the openings *x* leading into its upper and lower neighbors; *d*, tracheid; *e*, wood fibre proper.

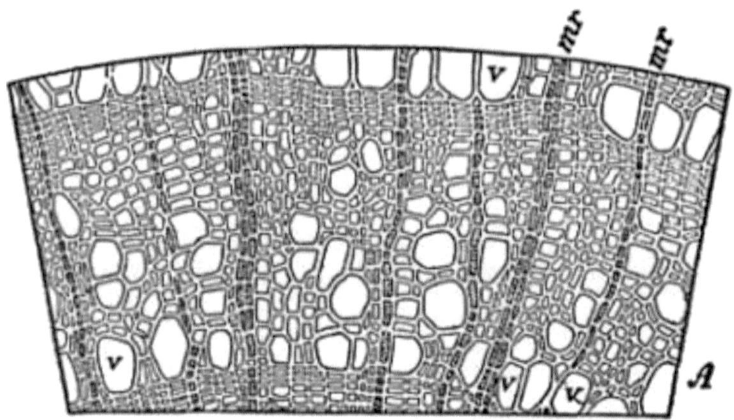

Fig. 9. Cross-section of Basswood (Magnified). *v*, vessels; *mr*, pith rays.

If a well-smoothed thin disk or cross-section of oak (say one-sixteenth inch thick) is held up to the light, it looks very much like a sieve, the pores or vessels appearing as clean-cut holes. The spring-wood and gray patches are seen to be quite porous, but the firm bodies of fibres between them are dense and opaque. Examined with a magnifier it will be noticed that there is no such regularity of arrangement in straight rows as is conspicuous in pine. On the contrary, great irregularity prevails. At the same time, while the pores [35] are as large as pin holes, the cells of the denser wood, unlike those of pine wood, are too small to be distinguished. Studied with the microscope, each vessel is found to be a vertical row of a great number of short, wide tubes, joined end to end (see Fig. 8, *c*). The porous spring-wood and radial gray tracts are partly composed of smaller vessels, but chiefly of tracheids, like those of pine, and of shorter cells, the "wood parenchyma," resembling the cells of the medullary rays. These latter, as well as the fine concentric lines mentioned as occurring in the summer-wood, are composed entirely of short tube-like parenchyma cells, with square or oblique ends (see Fig. 8, *a* and *b*). The wood fibres proper, which form the dark, firm bodies referred to, are very fine, thread-like cells, one twenty-fifth to one-tenth inch long, with a wall commonly so thick that scarcely any empty internal space or lumen remains (see Figs. 8, *e*, and 7, B). If, instead of oak, a piece of poplar or basswood (see Fig. 9) had been used in this study, the structure would have been found to be quite different. The same kinds of cell-elements, vessels, etc., are, to be sure, present, but their combination and arrangement are different, and thus from the great variety of possible combinations results the great variety of structure and, in consequence, of the qualities which distinguish the wood of broad-leaved trees. The sharp distinction of sap wood and heartwood is wanting; the rings are not so clearly defined; the vessels [36] of the wood are small, very numerous, and rather evenly scattered through the wood of the annual rings, so that the distinction of the ring almost vanishes and the medullary or pith rays in poplar can be seen, without being magnified, only on the radial section.

LIST OF MOST IMPORTANT BROAD-LEAVED TREES (HARDWOODS) [37]

Woods of complex and very variable structure, and therefore differing widely in quality, behavior, and consequently in applicability to the arts.

AILANTHUS

1. Ailanthus (*Ailanthus glandulosa*). Medium to large-sized tree. Wood pale yellow, hard, fine-grained, and satiny. This species originally came from China, where it is known as the Tree of "Heaven," was introduced into the United States and planted near Philadelphia during the 18th century, and is more ornamental than useful. It is used to some extent in cabinet work. Western Pennsylvania and Long Island, New York.

ASH

Wood heavy, hard, stiff, quite tough, not durable in contact with the soil, straight-grained, rough on the split surfaces and coarse in texture. The wood shrinks moderately, seasons with little injury, stands well, and takes a good polish. In carpentry, ash is used for stairways, panels, etc. It is used in shipbuilding, in the construction of cars, wagons, etc., in the manufacture of all kinds of farm implements, machinery, and especially of all kinds of furniture; for cooperage, baskets, oars, tool handles, hoops, etc., etc. The trees of the several species of ash are rapid growers, of small to medium height with stout trunks. They form no forests, but occur scattered in almost all our broad-leaved forests.

2. White Ash (*Fraxinus Americana*). Medium-, sometimes large-sized tree. Heartwood reddish brown, usually mottled; sapwood lighter color, nearly white. Wood heavy, hard, tough, elastic, coarse-grained, [38] compact structure. Annual rings clearly marked by large open pores, not durable in contact with the soil, is straight-grained, and the best material for oars, etc. Used for agricultural implements, tool handles, automobile (rim boards), vehicle bodies and parts, baseball bats, interior finish, cabinet work, etc., etc. Basin of the Ohio, but found from Maine to Minnesota and Texas.

3. Red Ash (*Fraxinus pubescens* var. *Pennsylvanica*). Medium-sized tree, a timber very similar to, but smaller than *Fraxinus Americana*. Heartwood light brown, sapwood lighter color. Wood heavy, hard, strong, and coarse-grained. Ranges from New Brunswick to Florida, and westward to Dakota, Nebraska, and Kansas.

4. Black Ash (*Fraxinus nigra* var. *sambucifolia*) (Hoop Ash, Ground Ash). Medium-sized tree, very common, is more widely distributed than the *Fraxinus Americana;* the wood is not so hard, but is well suited for hoops and basketwork. Heartwood dark brown, sapwood light brown or white. Wood heavy, rather soft, tough and coarse-grained. Used for barrel hoops, basketwork, cabinetwork and interior of houses. Maine to Minnesota and southward to Alabama.

5. Blue Ash (*Fraxinus quadrangulata*). Small to medium-sized tree. Heartwood yellow, streaked with brown, sapwood a lighter color. Wood heavy, hard, and coarse-grained. Not common. Indiana and Illinois; occurs from Michigan to Minnesota and southward to Alabama.

6. Green Ash (*Fraxinus viridis*). Small-sized tree. Occurs from New York to the Rocky Mountains, and southward to Florida and Arizona.

7. Oregon Ash (*Fraxinus Oregana*). Small to medium-sized tree. Occurs from western Washington to California. [39]

8. Carolina Ash (*Fraxinus Caroliniana*). Medium-sized tree. Occurs in the Carolinas and the coast regions southward.

ASPEN (See Poplar)

BASSWOOD

9. Basswood (*Tilia Americana*) (Linden, Lime Tree, American Linden, Lin, Bee Tree). Medium- to large-sized tree. Wood light, soft, stiff, but not strong, of fine texture, straight and close-grained, and white to light brown color, but not durable in contact with the soil. The wood shrinks considerably in drying, works well and stands well in interior work. It is used for cooperage, in carpentry, in the manufacture of furniture and woodenware (both turned and carved), for toys, also for panelling of car and carriage bodies, for agricultural implements, automobiles, sides and backs of drawers, cigar boxes, excelsior, refrigerators, trunks, and paper pulp. It is also largely cut for veneer and used as "three-ply" for boxes and chair seats. It is used for sounding boards in pianos and organs. If well seasoned and painted it stands fairly well for outside work. Common in all northern broad-leaved forests. Found throughout the eastern United States, but reaches its greatest size in the Valley of the Ohio, becoming often 130 feet in height, but its usual height is about 70 feet.

10. White Basswood (*Tilia heterophylla*) (Whitewood). A small-sized tree. Wood in its quality and uses similar to the preceding, only it is lighter in color. Most abundant in the Alleghany region.

11. White Basswood (*Tilia pubescens*) (Downy Linden, Small-leaved Basswood). Small-sized tree. Wood in its quality and uses similar to *Tilia Americana*. This is a Southern species which makes it way as far north as Long Island. Is found at its best in South Carolina. [40]

BEECH

12. Beech (*Fagus ferruginea*) (Red Beech, White Beech). Medium-sized tree, common, sometimes forming forests of pure growth. Wood heavy, hard, stiff, strong, of rather coarse texture, white to light brown color, not durable in contact with the soil, and subject to the inroads of boring insects. Rather close-grained, conspicuous medullary rays, and when quarter-sawn and well smoothed is very beautiful. The wood shrinks and checks considerably in drying, works well and stands well, and takes a fine polish. Beech is comparatively free from objectionable taste, and finds a place in the manufacture of commodities which come in contact with foodstuffs, such as lard tubs, butter boxes and pails, and the beaters of ice cream freezers; for the latter the persistent hardness of the wood when subjected to attrition and abrasion, while wet gives it peculiar fitness. It is an excellent material for churns. Sugar hogsheads are made of beech, partly because it is a tasteless wood and partly because it has great strength. A large class of woodenware, including veneer plates, dishes, boxes, paddles, scoops, spoons, and beaters, which belong to the kitchen and pantry, are made of this species of wood. Beech picnic plates are made by the million, a single machine turning out 75,000 a day. The wood has a long list of miscellaneous uses and enters in a great variety of commodities. In every region where it grows in commercial quantities it is made into boxes, baskets, and crating. Beech baskets are chiefly employed in shipping fruit, berries, and vegetables. In Maine thin veneer of beech is made specially for the Sicily orange and lemon trade. This is shipped in bulk and the boxes are made abroad. Beech is also an important handle wood, although not in the same class with hickory. It is not selected because of toughness and resiliency, as hickory is, and generally goes into plane, handsaw, pail, chisel, [41] and flatiron handles. Recent statistics show that in the production of slack cooperage staves, only two woods, red gum and pine, stood above beech in quantity, while for heading, pine alone exceeded it. It is also used in turnery, for shoe lasts, butcher blocks, ladder rounds, etc. Abroad it is very extensively used by the carpenter, millwright, and wagon maker, in turnery and wood carving. Most abundant in the Ohio and Mississippi basin, but found from Maine to Wisconsin and southward to Florida.

BIRCH

13. Cherry Birch (*Betula lenta*) (Black Birch, Sweet Birch, Mahogany Birch, Wintergreen Birch). Medium-sized tree, very common. Wood of beautiful reddish or yellowish brown, and much of it nicely figured, of compact structure, is straight in grain, heavy, hard, strong, takes a fine polish, and considerably used as imitation of mahogany. The wood shrinks considerably in drying, works well and stands well, but is not durable in contact with the soil. The medullary rays in birch are very fine and close and not easily seen. The sweet birch is very handsome, with satiny luster, equalling cherry, and is too costly a wood to be profitably used for ordinary purposes, but there are both high and low grades of birch, the latter consisting chiefly of sapwood and pieces too knotty for first class commodities. This cheap material swells the supply of box lumber, and a little of it is found wherever birch passes through sawmills. The frequent objections against sweet birch as box lumber and crating material are that it is hard to nail and is inclined to split. It is also used for veneer picnic plates and butter dishes, although it is not as popular for this class of commodity as are yellow and paper birch, maple and beech. The best grades are largely used for furniture and cabinet work, and also for interior finish. Maine to Michigan and to Tennessee. [42]

14. White Birch (*Betula populifolia*) (Gray Birch, Old Field Birch, Aspen-leaved Birch). Small to medium-sized tree, least common of all the birches. Short-lived, twenty to thirty feet high, grows very rapidly. Heartwood light brown, sapwood lighter color. Wood light, soft, close-grained, not strong, checks badly in drying, decays quickly, not durable in contact with the soil, takes a good polish. Used for spools, shoepegs, wood pulp, and barrel hoops. Fuel, value not high, but burns with bright flame. Ranges from Nova Scotia and lower St. Lawrence River, southward, mostly in the coast region to Delaware, and westward through northern New England and New York to southern shore of Lake Ontario.

15. Yellow Birch (*Betula lutea*) (Gray Birch, Silver Birch). Medium- to large-sized tree, very common. Heartwood light reddish brown, sapwood nearly white, close-grained, compact structure, with a satiny luster. Wood heavy, very strong, hard, tough, susceptible of high polish, not durable when exposed. Is similar to *Betula lenta*, and finds a place in practically all kinds of woodenware. A large percentage of broom handles on the market are made of this species of wood, though nearly every other birch contributes something. It is used for veneer plates and dishes made for pies, butter, lard, and many other commodities. Tubs and pails are sometimes made of yellow birch provided weight is not objectionable. The wood is twice as heavy as some of the pines and cedars. Many small handles for such articles as flatirons, gimlets, augers, screw drivers, chisels, varnish and paint brushes, butcher and carving knives, etc. It is also widely used for shipping boxes, baskets, and crates, and it is one of the stiffest, strongest woods procurable, but on account of its excessive weight it is sometimes discriminated against. It is excellent for veneer boxes, and that is probably one of the most important places it fills. Citrus fruit from northern Africa and the islands and countries of the Mediterranean is often shipped to market [43] in boxes made of yellow birch from veneer cut in New England. The better grades are also used for furniture and cabinet work, and the "burls" found on this species are highly valued for making fancy articles, gavels, etc. It is extensively used for turnery, buttons, spools, bobbins, wheel hubs, etc. Maine to Minnesota and southward to Tennessee.

16. Red Birch (*Betula rubra* var. *nigra*) (River Birch). Small to medium-sized tree, very common. Lighter and less valuable than the preceding. Heartwood light brown, sapwood pale. Wood light, fairly strong and close-grained. Red birch is best developed in the middle South, and usually grows near the banks of rivers. Its bark hangs in tatters, even worse than that of paper birch, but it is darker. In Tennessee the slack coopers have found that red birch makes excellent barrel heads and it is sometimes employed in preference to other woods. In eastern Maryland the manufacturers of peach baskets draw their supplies from this wood, and substitute it for white elm in making the hoops or bands which stiffen the top of the basket, and provide a fastening for the veneer which forms the sides. Red birch bends in a very satisfactory manner, which is an important point. This wood enters pretty generally into the manufacture of woodenware within its range, but statistics do not mention it by name. It is also used in the manufacture of veneer picnic plates, pie plates, butter dishes, washboards, small handles, kitchen and pantry utensils, and ironing boards. New England to Texas and Missouri.

17. Canoe Birch (*Betula paprifera*) (White Birch, Paper Birch). Small to medium-sized tree, sometimes forming forests, very common. Heartwood light brown tinged with red, sapwood lighter color. Wood of good quality, but light, fairly hard and strong, tough, close-grained. Sap flows freely in spring and by boiling can be made into syrup. Not as valuable as any of the preceding. Canoe birch is a northern [44] tree, easily identified by its white trunk and its ragged bark. Large numbers of small wooden boxes are made by boring out blocks of this wood, shaping them in lathes, and fitting lids on them. Canoe birch is one of the best woods for this class of commodities, because it can be worked very thin, does not split readily, and is of pleasing color. Such boxes, or two-piece diminutive kegs, are used as containers for articles shipped and sold in small bulk, such as tacks, small nails, and brads. Such containers are generally cylindrical and of considerably greater depth than diameter. Many others of nearly similar form are made to contain ink bottles, bottles of perfumery, drugs, liquids, salves, lotions, and powders of many kinds. Many boxes of this pattern are used by manufacturers of pencils and crayons for packing and shipping their wares. Such boxes are made in numerous numbers by automatic machinery. A single machine of the most improved pattern will turn out 1,400 boxes an hour. After the boring and turning are done, they are smoothed by placing them into a tumbling barrel with soapstone. It is also used for one-piece shallow trays or boxes, without lids, and used as card receivers, pin receptacles, butter boxes, fruit platters, and contribution plates in churches. It is also the principal wood used for spools, bobbins, bowls, shoe lasts, pegs, and turnery, and is also much used in the furniture trade. All along the northern boundary of the United States and northward, from the Atlantic to the Pacific.

BLACK WALNUT (See Walnut)

BLUE BEECH

18. Blue Beech (*Carpinus Caroliniana*) (Hornbeam, Water Beech, Ironwood). Small-sized tree. Heartwood light brown, sapwood nearly white. Wood very hard, heavy, strong, very stiff, of rather fine texture, not durable in contact with the soil, shrinks and checks considerably in drying, but works well and stands [45] well, and takes a fine polish. Used chiefly in turnery, for tool handles, etc. Abroad much used by mill- and wheelwrights. A small tree, largest in the Southwest, but found in nearly all parts of the eastern United States.

BOIS D'ARC (See Osage Orange)

BUCKEYE

Wood light, soft, not strong, often quite tough, of fine, uniform texture and creamy white color. It shrinks considerably in drying, but works well and stands well. Used for woodenware, artificial limbs, paper pulp, and locally also for building construction.

19. Ohio Buckeye (*Æsculus glabra*) (Horse Chestnut, Fetid Buckeye). Small-sized tree, scattered, never forming forests. Heartwood white, sapwood pale brown. Wood light, soft, not strong, often quite tough and close-grained. Alleghanies, Pennsylvania to Oklahoma.

20. Sweet Buckeye (*Æsculus octandra* var. *flava*) (Horse Chestnut). Small-sized tree, scattered, never forming forests. Wood in its quality and uses similar to the preceding. Alleghanies, Pennsylvania to Texas.

BUCKTHORNE

21. Buckthorne (*Rhanmus Caroliniana*) (Indian Cherry). Small-sized tree. Heartwood light brown, sapwood almost white. Wood light, hard, close-grained. Does not enter the markets to any great extent. Found along the borders of streams in rich bottom lands. Its northern limits is Long Island, where it is only a shrub; it becomes a tree only in southern Arkansas and adjoining regions.

BUTTERNUT

22. Butternut (*Juglans cinerea*) (White Walnut, White Mahogany, Walnut). Medium-sized tree, scattered, [46] never forming forests. Wood very similar to black walnut, but light, quite soft, and not strong. Heartwood light gray-brown, darkening with exposure; sapwood nearly white, coarse-grained, compact structure, easily worked, and susceptible to high polish. Has similar grain to black walnut and when stained is a very good imitation. Is much used for inside work, and very durable. Used chiefly for finishing lumber, cabinet work, boat finish and fixtures, and for furniture. Butternut furniture is often sold as circassian walnut. Largest and most common in the Ohio basin. Maine to Minnesota and southward to Georgia and Alabama.

CATALPA

The catalpa is a tree which was planted about 25 years ago as a commercial speculation in Iowa, Kansas, and Nebraska. Its native habitat was along the rivers Ohio and lower Wabash, and a century ago it gained a reputation for rapid growth and durability, but did not grow in large quantities. As a railway tie, experiments have left no doubt as to its resistance to decay; it stands abrasion as well as the white oak (*Quercus alba*), and is superior to it in longevity. Catalpa is a tree singularly free from destructive diseases. Wood cut from the living tree is one of the most durable timbers known. In spite of its light porous structure it resists the weathering influences and the attacks of wood-destroying fungi to a remarkable degree. No fungus has yet been found which will grow in the dead timber, and for fence posts this wood has no equal, lasting longer than almost any other species of timber. The wood is rather soft and coarse in texture, the tree is of slow growth, and the brown colored heartwood, even of very young trees, forms nearly three-quarters of their volume. There is only about one-quarter inch of sapwood in a 9-inch tree.

23. Catalpa (*Catalpa speciosa* var. *bignonioides*) (Indian Bean). Medium-sized tree. Heartwood light brown, sapwood nearly white. Wood light, soft, not strong, [47] brittle, very durable in contact with the soil, of coarse texture. Used chiefly for railway ties, telegraph poles, and fence posts, but well suited for a great variety of uses. Lower basin of the Ohio River, locally common. Extensively planted, and therefore promising to become of some importance.

CHERRY

24. Cherry (*Prunus serotina*) (Wild Cherry, Black Cherry, Rum Cherry). Wood heavy, hard, strong, of fine texture. Sapwood yellowish white, heartwood reddish to brown. The wood shrinks considerably in drying, works well and stands well, has a fine satin-like luster, and takes a fine polish which somewhat resembles mahogany, and is much esteemed for its beauty. Cherry is chiefly used as a decorative interior finishing lumber, for buildings, cars and boats, also for furniture and in turnery, for musical instruments, walking sticks, last blocks, and woodenware. It is becoming too costly for many purposes for which it is naturally well suited. The lumber-furnishing cherry of the United States, the wild black cherry, is a small to medium-sized tree, scattered through many of the broad-leaved trees of the western slope of the Alleghanies, but found from Michigan to Florida, and west to Texas. Other species of this genus, as well as the hawthornes (*Prunus cratoegus*) and wild apple (*Pyrus*), are not commonly offered in the markets. Their wood is of the same character as cherry, often finer, but in smaller dimensions.

25. Red Cherry (*Prunus Pennsylvanica*) (Wild Red Cherry, Bird Cherry). Small-sized tree. Heartwood light brown, sapwood pale yellow. Wood light, soft, and close-grained. Uses similiar to the preceding, common throughout the Northern States, reaching its greatest size on the mountains of Tennessee. [48]

CHESTNUT

The chestnut is a long-lived tree, attaining an age of from 400 to 600 years, but trees over 100 years are usually hollow. It grows quickly, and sprouts from a chestnut stump (Coppice Chestnut) often attain a height of 8 feet in the first year. It has a fairly cylindrical stem, and often grows to a height of 100 feet and over. Coppice chestnut, that is, chestnut grown on an old stump, furnishes better timber for working than chestnut grown from the nut, it is heavier, less spongy, straighter in grain, easier to split, and stands exposure longer.

26. Chestnut (*Castanea vulgaris* var. *Americana*). Medium- to large-sized tree, never forming forests. Wood is light, moderately hard, stiff, elastic, not strong, but very durable when in contact with the soil, of coarse texture. Sapwood light, heartwood darker brown, and is readily distinguishable from the sapwood, which very early turns into heartwood. It shrinks and checks considerably in drying, works easily, stands well. The annual rings are very distinct, medullary rays very minute and not visible to the naked eye. Used in cooperage, for cabinetwork, agricultural implements, railway ties, telegraph poles, fence posts, sills, boxes, crates, coffins, furniture, fixtures, foundation for veneer, and locally in heavy construction. Very common in the Alleghanies. Occurs from Maine to Michigan and southward to Alabama.

27. Chestnut (*Castanea dentata* var. *vesca*). Medium-sized tree, never forming forests, not common. Heartwood brown color, sapwood lighter shade, coarse-grained. Wood and uses similar to the preceding. Occurs scattered along the St. Lawrence River, and even there is met with only in small quantities.

28. Chinquapin (*Castanea pumila*). Medium- to small-sized tree, with wood slightly heavier, but otherwise similiar to the preceding.

Most common in Arkansas, but with nearly the same range as *Castanea vulgaris*. [49]

29. Chinquapin (*Castanea chrysophylla*). A medium-sized tree of the western ranges of California and Oregon.

COFFEE TREE

30. Coffee Tree (*Gymnocladus dioicus*) (Coffee Nut, Stump Tree). A medium- to large-sized tree, not common. Wood heavy, hard, strong, very stiff, of coarse texture, and durable. Sapwood yellow, heartwood reddish brown, shrinks and checks considerably in drying, works well and stands well, and takes a fine polish. It is used to a limited extent in cabinetwork and interior finish. Pennsylvania to Minnesota and Arkansas.

COTTONWOOD (See Poplar)

CRAB APPLE

31. Crab Apple (*Pyrus coronaria*) (Wild Apple, Fragrant Crab). Small-sized tree. Heartwood reddish brown, sapwood yellow. Wood heavy, hard, not strong, close-grained. Used principally for tool handles and small domestic articles. Most abundant in the middle and western states, reaches its greatest size in the valleys of the lower Ohio basin.

CUCUMBER TREE (See Magnolia)

DOGWOOD

32. Dogwood (*Cornus florida*) (American Box). Small to medium-sized tree. Attains a height of about 30 feet and about 12 inches in diameter. The heartwood is a red or pinkish color, the sapwood, which is considerable, is a creamy white. The wood has a dull surface and very fine grain. It is valuable for turnery, tool handles, and mallets, and being so free from silex, watchmakers use small splinters of it for cleaning out the pivot holes of watches, and opticians for removing dust from deep-seated lenses. It is [50] also used for butchers' skewers, and shuttle blocks and wheel stock, and is suitable for turnery and inlaid work. Occurs scattered in all the broad-leaved forests of our country; very common.

ELM

Wood heavy, hard, strong, elastic, very tough, moderately durable in contact with the soil, commonly cross-grained, difficult to split and shape, warps and checks considerably in drying, but stands well if properly seasoned. The broad sapwood whitish, heartwood light brown, both with shades of gray and red. On split surfaces rough, texture coarse to fine, capable of high polish. Elm for years has been the principal wood used in slack cooperage for barrel staves, also in the construction of cars, wagons, etc., in boat building, agricultural implements and machinery, in saddlery and harness work, and particularly in the manufacture of all kinds of furniture, where the beautiful figures, especially those of the tangential or bastard section, are just beginning to be appreciated. The elms are medium- to large-sized trees, of fairly rapid growth, with stout trunks; they form no forests of pure growth, but are found scattered in all the broad-leaved woods of our country, sometimes forming a considerable portion of the arborescent growth.

33. White Elm (*Ulmus Americana*) (American Elm, Water Elm). Medium- to large-sized tree. Wood in its quality and uses as stated above. Common. Maine to Minnesota, southward to Florida and Texas.

34. Rock Elm (*Ulmus racemosa*) (Cork Elm, Hickory Elm, White Elm, Cliff Elm). Medium- to large-sized tree of rapid growth. Heartwood light brown, often tinged with red, sapwood yellowish or greenish white, compact structure, fibres interlaced. Wood heavy, hard, very tough, strong, elastic, difficult to split, takes a fine polish. Used for agricultural implements, automobiles, crating, boxes, cooperage, tool handles, wheel stock, bridge timbers, sills, interior [51] finish, and maul heads. Fairly free from knots and has only a small quantity of sapwood. Michigan, Ohio, from Vermont to Iowa, and southward to Kentucky.

35. Red Elm (*Ulmus fulva* var. *pubescens*) (Slippery Elm, Moose Elm). The red or slippery elm is not as large a tree as the white elm (*Ulmus Americana*), though it occasionally attains a height of 135 feet and a diameter of 4 feet. It grows tall and straight, and thrives in river valleys. The wood is heavy, hard, strong, tough, elastic, commonly cross-grained, moderately durable in contact with the soil, splits easily when green, works fairly well, and stands well if properly handled. Careful seasoning and handling are essential for the best results. Trees can be utilized for posts when very small. When green the wood rots very quickly in contact with the soil. Poles for posts should be cut in summer and peeled and dried before setting. The wood becomes very tough and pliable when steamed, and is of value for sleigh runners and for ribs of canoes and skiffs. Together with white elm (*Ulmus Americana*) it is extensively used for barrel staves in slack cooperage and also for furniture. The thick, viscous inner bark, which gives the tree its descriptive name, is quite palatable, slightly nutritious, and has a medicinal value. Found chiefly along water courses. New York to Minnesota, and southward to Florida and Texas.

36. Cedar Elm (*Ulmus crassifolia*). Medium- to small-sized tree, locally quite common. Arkansas and Texas.

37. Winged Elm (*Ulmus alata*) (Wahoo). Small-sized tree, locally quite common. Heartwood light brown, sapwood yellowish white. Wood heavy, hard, tough, strong, and close-grained. Arkansas, Missouri, and eastern Virginia. [52]

Fig. 10. A Large Red Gum.

GUM

This general term applies to three important species of gum in the South, the principal one usually being distinguished as "red" or "sweet" gum (see Fig. 10). The next in importance being the "tupelo" or "bay poplar," and the least of the trio is designated as "black" or "sour" gum (see Fig. 11). Up to the year 1900 little was known of gum as a wood for cooperage purposes, but [53] by the continued advance in price of the woods used, a few of the most progressive manufacturers, looking into the future, saw that the supply of the various woods in use was limited, that new woods would have to be sought, and gum was looked upon as a possible substitute, owing to its cheapness and abundant supply. No doubt in the future this wood will be used to a considerable extent in the manufacture of both "tight" and "slack" cooperage. [54] In the manufacture of the gum, unless the knives and saws are kept very sharp, the wood has a tendency to break out, the corners splitting off; and also, much difficulty has been experienced in seasoning and kiln-drying.

Fig. 11. A Tupelo Gum Slough.

In the past, gum, having no marketable value, has been left standing after logging operations, or, where the land has been cleared for farming, the trees have been "girdled" and allowed to rot, and then felled and burned as trash. Now, however, that there is a market for this species of timber, it will be profitable to cut the gum with the other hardwoods, and this species of wood will come in for a greater share of attention than ever before.

38. Red Gum (*Liquidamber styraciflua*) (Sweet Gum, Hazel Pine, Satin Walnut, Liquidamber, Bilsted). The wood is about as stiff and as strong as chestnut, rather heavy, it splits easily and is quite brash, commonly cross-grained, of fine texture, and has a large proportion of whitish sapwood, which decays rapidly when exposed to the weather; but the reddish brown heartwood is quite durable, even in the ground. The external appearance of the wood is of fine grain and smooth, close texture, but when broken the lines of fracture do not run with apparent direction of the growth; possibly it is this unevenness of grain which renders the wood so difficult to dry without twisting and warping. It has little resiliency; can be easily bent when steamed, and when properly dried will hold its shape. The annual rings are not distinctly marked, medullary rays fine and numerous. The green wood contains much water, and consequently is heavy and difficult to float, but when dry it is as light as basswood. The great amount of water in the green wood, particularly in the sap, makes it difficult to season by ordinary methods without warping and twisting. It does not check badly, is tasteless and odorless, and when once seasoned, swells and shrinks but little unless exposed to the weather. Used for boat finish, veneers, cabinet work, furniture, fixtures, interior decoration, shingles, paving blocks, woodenware, [55] cooperage, machinery frames, refrigerators, and trunk slats.

Range of Red Gum

Red gum is distributed from Fairfield County, Conn., to southeastern Missouri, through Arkansas and Oklahoma to the valley of the Trinity River in Texas, and eastward to the Atlantic coast. Its commercial range is restricted, however, to the moist lands of the lower Ohio and Mississippi basins and of the Southeastern coast. It is one of the commonest timber trees in the hardwood bottoms and drier swamps of the South. It grows in mixture with ash, cottonwood and oak (see Fig. 12). It is also found to a considerable extent on the lower ridges and slopes of the southern Appalachians, but there it does not reach merchantable value and is of little importance. Considerable difference is found between the growth in the upper Mississippi bottoms and that along the rivers on the Atlantic coast and on the Gulf. In the latter regions the bottoms are lower, and consequently more subject to floods and to continued overflows (see Fig. 11). The alluvial deposit is also greater, and the trees grow considerably faster. Trees of the same diameter show a larger percentage of sapwood there than in the upper portions of the Mississippi Valley. The Mississippi Valley hardwood trees are for the most part considerably older, and reach larger dimensions than the timber along the coast.

Form of the Red Gum

In the best situations red gum reaches a height of 150 feet, and a diameter of 5 feet. These dimensions, however are unusual. The stem is straight and cylindrical, with dark, deeply-furrowed bark, and branches often winged with corky ridges. In youth, while growing vigorously under normal conditions, it assumes a long, regular, conical crown, much resembling the form of a conifer (see Fig. 12). After the tree has attained its height growth, however, the crown becomes rounded, spreading and rather ovate in shape. When growing in the forest [56] the tree prunes itself readily at an early period, and forms a good length of clear stem, but it branches strongly after making most of its height growth. The mature tree is usually forked, and the place where the forking commences determines the number of logs in the tree or its merchantable length, by preventing cutting to a small diameter in the top. On large trees the stem is often not less than eighteen inches in diameter where the branching begins. The over-mature tree is usually broken and dry topped, with a very spreading crown, in consequence of new branches being sent out.

Tolerance of Red Gum

Throughout its entire life red gum is intolerant in shade, there are practically no red seedlings under the dense forest cover of the bottom land, and while a good many may come up under the pine forest on the drier uplands, they seldom develop into large trees. As a rule seedlings appear only in clearings or in open spots in the forest. It is seldom that an over-topped tree is found, for the gum dies quickly if suppressed, and is consequently nearly always a dominant or intermediate tree. In a hardwood bottom forest the timber trees are all of nearly the same age over considerable areas, and there is little young growth to be found in the older stands. The reason for this is the intolerance of most of the swamp species. A scale of intolerance containing the important species, and beginning with the most light-demanding, would run as follows: Cottonwood, sycamore, red gum, white elm, white ash, and red maple.

Demands upon Soil and Moisture

While the red gum grows in various situations, it prefers the deep, rich soil of the hardwood bottoms, and there reaches its best development (see Fig. 10). It requires considerable soil moisture, though it does not grow in the wetter swamps, and does not thrive on dry pine land. Seedlings, however, are often found in large numbers on the edges of the uplands and even on the sandy pine land, but they seldom live beyond the pole stage. When they [57] do, they form small, scrubby trees that are of little value. Where the soil is dry the tree has a long tap root. In the swamps, where the roots can obtain water easily, the development of the tap root is poor, and it is only moderate on the glade bottom lands, where there is considerable moisture throughout the year, but no standing water in the summer months.

Fig. 12. Second Growth Red Gum, Ash, Cottonwood, and Sycamore.

Red gum reproduces both by seed and by sprouts (see Fig. 12). It produces seed fairly abundantly every year, but about once in three years there is an extremely heavy production. The tree begins to bear seed when twenty-five to thirty years old, and seeds vigorously up to an age of one hundred and fifty years, when its productive power begins to diminish. A great part of the seed, however, is abortive. Red gum is not fastidious in regard to its germinating bed; it comes up readily on sod [58] in old fields and meadows, on decomposing humus in the forest, or on bare clay-loam or loamy sand soil. It requires a considerable degree of light, however, and prefers a moist seed bed. The natural distribution of the seed takes place for several hundred feet from the seed trees, the dissemination depending almost entirely on the wind. A great part of the seed falls on the hardwood bottom when the land is flooded, and is either washed away or, if already [59] in the ground and germinating, is destroyed by the long-continued overflow. After germinating, the red gum seedling demands, above everything else, abundant light for its survival and development. It is for this reason that there is very little growth of red gum, either in the unculled forest or on culled land, where, as is usually the case, a dense undergrowth of cane, briers, and rattan is present. Under the dense underbrush of cane and briers throughout much of the virgin forest, reproduction of any of the merchantable species is of course impossible. And even where the land has been logged over, the forest is seldom open enough to allow reproduction of cottonwood and red gum. Where, however, seed trees are contiguous to pastures or cleared land, scattered seedlings are found springing up in the open, and where openings occur in the forest, there are often large numbers of red gum seedlings, the reproduction generally occurring in groups. But over the greater part of the Southern hardwood bottom land forest reproduction is very poor. The growth of red gum during the early part of its life, and up to the time it reaches a diameter of eight inches breast-high, is extremely rapid, and, like most of the intolerant species, it attains its height growth at an early period. Gum sprouts readily from the stump, and the sprouts surpass the seedlings in rate of height growth for the first few years, but they seldom form large timber trees. Those over fifty years of age seldom sprout. For this reason sprout reproduction is of little importance in the forest. The principal requirements of red gum, then, are a moist, fairly rich soil and good exposure to light. Without these it will not reach its best development.

Fig. 13. A Cypress Slough in the Dry Season.

Second-Growth Red Gum

Second-growth red gum occurs to any considerable extent only on land which has been thoroughly cleared. Throughout the South there is a great deal of land which was in cultivation before the Civil War, but which during the subsequent period of industrial depression was abandoned and allowed to revert to forest. These old fields now mostly covered with second-growth forest, of [60] which red gum forms an important part (see Fig. 12). Frequently over fifty per cent of the stand consists of this species, but more often, and especially on the Atlantic coast, the greater part is of cottonwood or ash. These stands are very dense, and the growth is extremely rapid. Small stands of young growth are also often found along the edges of cultivated fields. In the Mississippi Valley the abandoned fields on which young stands have sprung up are for the most part being rapidly cleared again. The second growth here is considered of little value in comparison with the value of the land for agricultural purposes. In many cases, however, the farm value of the land is not at present sufficient to make it profitable to clear it, unless the timber cut will at least pay for the operation. There is considerable land upon which the second growth will become valuable timber within a few years. Such land should not be cleared until it is possible to utilize the timber.

39. Tupelo Gum (*Nyssa aquatica*) (Bay Poplar, Swamp Poplar, Cotton Gum, Hazel Pine, Circassian Walnut, Pepperidge, Nyssa). The close similarity which exists between red and tupelo gum, together with the fact that tupelo is often cut along with red gum, and marketed with the sapwood of the latter, makes it not out of place to give consideration to this timber. The wood has a fine, uniform texture, is moderately hard and strong, is stiff, not elastic, very tough and hard to split, but easy to work with tools. Tupelo takes glue, paint, or varnish well, and absorbs very little of the material. In this respect it is equal to yellow poplar and superior to cottonwood. The wood is not durable in contact with ground, and requires much care in seasoning. The distinction between the heartwood and sapwood of this species is marked. The former varies in color from a dull gray to a dull brown; the latter is whitish or light yellow like that of poplar. The wood is of medium weight, about thirty-two pounds per cubic foot when dry, or nearly that of red gum and loblolly pine. After [61] seasoning it is difficult to distinguish the better grades of sapwood from poplar. Owing to the prejudice against tupelo gum, it was until recently marketed under such names as bay poplar, swamp poplar, nyssa, cotton gum, circassian walnut, and hazel pine. Since it has become evident that the properties of the wood fit it for many uses, the demand for tupelo has largely increased, and it is now taking rank with other standard woods under its rightful name. Heretofore the quality and usefulness of this wood were greatly underestimated, and the difficulty of handling it was magnified. Poor success in seasoning and kiln-drying was laid to defects of the wood itself, when, as a matter of fact, the failures were largely due to the absence of proper methods in handling. The passing of this prejudice against tupelo is due to a better understanding of the characteristics and uses of the wood. Handled in the way in which its particular character demands, tupelo is a wood of much value.

Uses of Tupelo Gum

Tupelo gum is now used in slack cooperage, principally for heading. It is used extensively for house flooring and inside finishing, such as mouldings, door jambs, and casings. A great deal is now shipped to European countries, where it is highly valued for different classes of manufacture. Much of the wood is used in the manufacture of boxes, since it works well upon rotary veneer machines. There is also an increasing demand for tupelo for laths, wooden pumps, violin and organ sounding boards, coffins, mantelwork, conduits and novelties. It is also used in the furniture trade for backing, drawers, and panels.

Range of Tupelo Gum

Tupelo occurs throughout the coastal region of the Atlantic States, from southern Virginia to northern Florida, through the Gulf States to the valley of the Nueces River in Texas, through Arkansas and southern Missouri to western Kentucky and Tennessee, and to the valley of [62] the lower Wabash River. Tupelo is being extensively milled at present only in the region adjacent to Mobile Ala., and in southern and central Louisiana, where it occurs in large merchantable quantities, attaining its best development in the former locality. The country in this locality is very swampy (see Fig. 11), and within a radius of one hundred miles tupelo gum is one of the principal timber trees. It grows only in the swamps and wetter situations (see Fig. 11), often in mixture with cypress, and in the rainy season it stands in from two to twenty feet of water.

40. Black Gum (*Nyssa sylvatica*) (Sour Gum). Black gum is not cut to much extent, owing to its less abundant supply and poorer quality, but is used for repair work on wagons, for boxes, crates, wagon hubs, rollers, bowls, woodenware, and for cattle yokes and other purposes which require a strong, non-splitting wood. Heartwood is light brown in color, often nearly white; sapwood hardly distinguishable, fine grain, fibres interwoven. Wood is heavy, not hard, difficult to work, strong, very tough, checks and warps considerably in drying, not durable. It is distributed from Maine to southern Ontario, through central Michigan to southeastern Missouri, southward to the valley of the Brazos River in Texas, and eastward to the Kissimmee River and Tampa Bay in Florida. It is found in the swamps and hardwood bottoms, but is more abundant and of better size on the slightly higher ridges and hummocks in these swamps, and on the mountain slopes in the southern Alleghany region. Though its range is greater than that of either red or tupelo gum, it nowhere forms an important part of the forest.

HACKBERRY

41. Hackberry (*Celtis occidentalis*) (Sugar Berry, Nettle Tree). The wood is handsome, heavy, hard, strong, quite tough, of moderately fine texture, and greenish or yellowish color, shrinks moderately, works well [63] and stands well, and takes a good polish. Used to some extent in cooperage, and in the manufacture of cheap furniture. Medium- to large-sized tree, locally quite common, largest in the lower Mississippi Valley. Occurs in nearly all parts of the eastern United States.

HICKORY

The hickories of commerce are exclusively North American and some of them are large and beautiful trees of 60 to 70 feet or more in height. They are closely allied to the walnut, and the wood is very like walnut in grain and color, though of a somewhat darker brown. It is one of the finest of American hardwoods in point of strength; in toughness it is superior to ash, rather coarse in texture, smooth and of straight grain, very heavy and strong as well as elastic and tenacious, but decays rapidly, especially the sapwood when exposed to damp and moisture, and is very liable to attack from worms and boring insects. The cross-section of hickory is peculiar, the annual rings appear like fine lines instead of like the usual pores, and the medullary rays, which are also very fine but distinct, in crossing these form a peculiar web-like pattern which is one of the characteristic differences between hickory and ash. Hickory is rarely subjected to artificial treatment, but there is this curious fact in connection with the wood, that, contrary to most other woods, creosote is only with difficulty injected into the sap, although there is no difficulty in getting it into the heartwood. It dries slowly, shrinks and checks considerably in seasoning; is not durable in contact with the soil or if exposed. Hickory excels as wagon and carriage stock, for hoops in cooperage, and is extensively used in the manufacture of implements and machinery, for tool handles, timber pins, harness work, dowel pins, golf clubs, and fishing rods. The hickories are tall trees with slender stems, never forming forests, occasionally small groves, but usually occur scattered among other broad-leaved trees in suitable localities. The following species all contribute more or less to the hickory of the markets: [64]

42. Shagbark Hickory (*Hicoria ovata*) (Shellbark Hickory, Scalybark Hickory). A medium- to large-sized tree, quite common; the favorite among the hickories. Heartwood light brown, sapwood ivory or cream-colored. Wood close-grained, compact structure, annual rings clearly marked. Very hard, heavy, strong, tough, and flexible, but not durable in contact with the soil or when exposed. Used for agricultural implements, wheel runners, tool handles, vehicle parts, baskets, dowel pins, harness work, golf clubs, fishing rods, etc. Best developed in the Ohio and Mississippi basins; from Lake Ontario to Texas, Minnesota to Florida.

43. Mockernut Hickory (*Hicoria alba*) (Black Nut Hickory, Black Hickory, Bull Nut Hickory, Big Bud Hickory, White Heart Hickory). A medium- to large-sized tree. Wood in its quality and uses similar to the preceding. Its range is the same as that of *Hicoria ovata*. Common, especially in the South.

44. Pignut Hickory (*Hicoria glabra*) (Brown Hickory, Black Hickory, Switchbud Hickory). A medium- to large-sized tree. Heavier and stronger than any of the preceding. Heartwood light to dark brown, sapwood nearly white. Abundant, all eastern United States.

45. Bitternut Hickory (*Hicoria minima*) (Swamp Hickory). A medium-sized tree, favoring wet localities. Heartwood light brown, sapwood lighter color. Wood in its quality and uses not so valuable as *Hicoria ovata*, but is used for the same purposes. Abundant, all eastern United States.

46. Pecan (*Hicoria pecan*) (Illinois Nut). A large tree, very common in the fertile bottoms of the western streams. Indiana to Nebraska and southward to Louisiana and Texas.

HOLLY

47. Holly (*Ilex opaca*). Small to medium-sized tree. Wood of medium weight, hard, strong, tough, of [65] exceedingly fine grain, closer in texture than most woods, of white color, sometimes almost as white as ivory; requires great care in its treatment to preserve the whiteness of the wood. It does not readily absorb foreign matter. Much used by turners and for all parts of musical instruments, for handles on whips and fancy articles, draught-boards, engraving blocks, cabinet work, etc. The wood is often dyed black and sold as ebony; works well and stands well. Most abundant in the lower Mississippi Valley and Gulf States, but occurring eastward to Massachusetts and north to Indiana.

48. Holly (*Ilex monticolo*) (Mountain Holly). Small-sized tree. Wood in its quality and uses similar to the preceding, but is not very generally known. It is found in the Catskill Mountains and extends southward along the Alleghanies as far as Alabama.

HORSE CHESTNUT (See Buckeye)

IRONWOOD

49. Ironwood (*Ostrya Virginiana*) (Hop Hornbeam, Lever Wood). Small-sized tree, common. Heartwood light brown tinged with red, sapwood nearly white. Wood heavy, tough, exceedingly close-grained, very strong and hard, durable in contact with the soil, and will take a fine polish. Used for small articles like levers, handles of tools, mallets, etc. Ranges throughout the United States east of the Rocky Mountains.

LAUREL

50. Laurel (*Umbellularia Californica*) (Myrtle). A Western tree, produces timber of light brown color of great size and beauty, and is very valuable for cabinet and inside work, as it takes a fine polish. California and Oregon, coast range of the Sierra Nevada Mountains. [66]

LOCUST

51. Black Locust (*Robinia pseudacacia*) (Locust, Yellow Locust, Acacia). Small to medium-sized tree. Wood very heavy, hard, strong, and tough, rivalling some of the best oak in this latter quality. The wood has great torsional strength, excelling most of the soft woods in this respect, of coarse texture, close-grained and compact structure, takes a fine polish. Annual rings clearly marked, very durable in contact with the soil, shrinks and checks considerably in drying, the very narrow sapwood greenish yellow, the heartwood brown, with shades of red and green. Used for wagon hubs, trenails or pins, but especially for railway ties, fence posts, and door sills. Also used for boat parts, turnery, ornamentations, and locally for construction. Abroad it is much used for furniture and farming implements and also in turnery. At home in the Alleghany Mountains, extensively planted, especially in the West.

52. Honey Locust (*Gleditschia triacanthos*) (Honey Shucks, Locust, Black Locust, Brown Locust, Sweet Locust, False Acacia, Three-Thorned Acacia). A medium-sized tree. Wood heavy, hard, strong, tough, durable in contact with the soil, of coarse texture, susceptible to a good polish. The narrow sapwood yellow, the heartwood brownish red. So far, but little appreciated except for fences and fuel. Used to some extent for wheel hubs, and locally in rough construction. Found from Pennsylvania to Nebraska, and southward to Florida and Texas; locally quite abundant.

53. Locust (*Robinia viscosa*) (Clammy Locust). Usually a shrub five or six feet high, but known to reach a height of 40 feet in the mountains of North Carolina, with the habit of a tree. Wood light brown, heavy, hard, and close-grained. Not used to much extent in manufacture. Range same as the preceding. [67]

MAGNOLIA

54. Magnolia (*Magnolia glauca*) (Swamp Magnolia, Small Magnolia, Sweet Bay, Beaver Wood). Small-sized tree. Heartwood reddish brown, sap wood cream white. Sparingly used in manufacture. Ranges from Essex County, Mass., to Long Island, N. Y., from New Jersey to Florida, and west in the Gulf region to Texas.

55. Magnolia (*Magnolia tripetala*) (Umbrella Tree). A small-sized tree. Wood in its quality similiar to the preceding. It may be easily recognized by its great leaves, twelve to eighteen inches long, and five to eight inches broad. This species as well as the preceding is an ornamental tree. Ranges from Pennsylvania southward to the Gulf.

56. Cucumber Tree (*Magnolia accuminata*) (Tulip-wood, Poplar). Medium- to large-sized tree. Heartwood yellowish brown, sapwood almost white. Wood light, soft, satiny, close-grained, durable in contact with the soil, resembling and sometimes confounded with tulip tree (*Liriodendron tulipifera*) in the markets. The wood shrinks considerably, but seasons without much injury, and works and stands well. It bends readily when steamed, and takes stain and paint well. Used in cooperage, for siding, for panelling and finishing lumber in house, car and shipbuilding, etc., also in the manufacture of toys, culinary woodenware, and backing for drawers. Most common in the southern Alleghanies, but distributed from western New York to southern Illinois, south through central Kentucky and Tennessee to Alabama, and throughout Arkansas.

MAPLE

Wood heavy, hard, strong, stiff, and tough, of fine texture, frequently wavy-grained, this giving rise to "curly" and "blister" figures which are much admired. Not durable in the ground, or when exposed. Maple [68] is creamy white, with shades of light brown in the heartwood, shrinks moderately, seasons, works, and stands well, wears smoothly, and takes a fine polish. The wood is used in cooperage, and for ceiling, flooring, panelling, stairway, and other finishing lumber in house, ship, and car construction. It is used for the keels of boats and ships, in the manufacture of implements and machinery, but especially for furniture, where entire chamber sets of maple rival those of oak. Maple is also used for shoe lasts and other form blocks; for shoe pegs; for piano actions, school apparatus, for wood type in show bill printing, tool handles, in wood carving, turnery, and scroll work, in fact it is one of our most useful woods. The maples are medium-sized trees, of fairly rapid growth, sometimes form forests, and frequently constitute a large proportion of the arborescent growth. They grow freely in parts of the Northern Hemisphere, and are particularly luxuriant in Canada and the northern portions of the United States.

57. Sugar Maple (*Acer saccharum*) (Hard Maple, Rock Maple). Medium- to large-sized tree, very common, forms considerable forests, and is especially esteemed. The wood is close-grained, heavy, fairly hard and strong, of compact structure. Heartwood brownish, sapwood lighter color; it can be worked to a satin-like surface and take a fine polish, it is not durable if exposed, and requires a good deal of seasoning. Medullary rays small but distinct. The "curly" or "wavy" varieties furnish wood of much beauty, the peculiar contortions of the grain called "bird's eye" being much sought after, and used as veneer for panelling, etc. It is used in all good grades of furniture, cabinetmaking, panelling, interior finish, and turnery; it is not liable to warp and twist. It is also largely used for flooring, for rollers for wringers and mangling machines, for which there is a large and increasing demand. The peculiarity known as "bird's eye," and which causes a difficulty in working the wood smooth, owing to the little pieces like knots [69] lifting up, is supposed to be due to the action of boring insects. Its resistance to compression across the grain is higher than that of most other woods. Ranges from Maine to Minnesota, abundant, with birch, in the region of the Great Lakes.

58. Red Maple (*Acer rubrum*) (Swamp Maple, Soft Maple, Water Maple). Medium-sized tree. Like the preceding but not so valuable. Scattered along water-courses and other moist localities. Abundant. Maine to Minnesota, southward to northern Florida.

59. Silver Maple (*Acer saccharinum*) (Soft Maple, White Maple, Silver-Leaved Maple). Medium- to large-sized tree, common. Wood lighter, softer, and inferior to *Acer saccharum,* and usually offered in small quantities and held separate in the markets. Heartwood reddish brown, sapwood ivory white, fine-grained, compact structure.

Fibres sometimes twisted, weaved, or curly. Not durable. Used in cooperage for woodenware, turnery articles, interior decorations and flooring. Valley of the Ohio, but occurs from Maine to Dakota and southward to Florida.

60. Broad-Leaved Maple (*Acer macrophyllum*) (Oregon Maple). Medium-sized tree, forms considerable forests, and, like the preceding has a lighter, softer, and less valuable wood than *Acer saccharum*. Pacific Coast regions.

61. Mountain Maple (*Acer spicatum*). Small-sized tree. Heartwood pale reddish brown, sapwood lighter color. Wood light, soft, close-grained, and susceptible of high polish. Ranges from lower St. Lawrence River to northern Minnesota and regions of the Saskatchewan River; south through the Northern States and along the Appalachian Mountains to Georgia.

62. Ash-Leaved Maple (*Acer negundo*) (Box Elder). Medium- to large-sized tree. Heartwood creamy white, sapwood nearly white. Wood light, soft, close-grained, [70] not strong. Used for woodenware and paper pulp. Distributed across the continent, abundant throughout the Mississippi Valley along banks of streams and borders of swamps.

63. Striped Maple (*Acer Pennsylvanicum*) (Moose-wood). Small-sized tree. Produces a very white wood much sought after for inlaid and for cabinet work. Wood is light, soft, close-grained, and takes a fine polish. Not common. Occurs from Pennsylvania to Minnesota.

MULBERRY

64. Red Mulberry (*Morus rubra*). A small-sized tree. Wood moderately heavy, fairly hard and strong, rather tough, of coarse texture, very durable in contact with the soil. The sapwood whitish, heartwood yellow to orange brown, shrinks and checks considerably in drying, works well and stands well. Used in cooperage and locally in construction, and in the manufacture of farm implements. Common in the Ohio and Mississippi Valleys, but widely distributed in the eastern United States.

MYRTLE (See Laurel)

OAK

Wood very variable, usually very heavy and hard, very strong and tough, porous, and of coarse texture. The sapwood whitish, the heartwood "oak" to reddish brown. It shrinks and checks badly, giving trouble in seasoning, but stands well, is durable, and little subject to the attacks of boring insects. Oak is used for many purposes, and is the chief wood used for tight cooperage; it is used in shipbuilding, for heavy construction, in carpentry, in furniture, car and wagon work, turnery, and even in woodcarving. It is also used in all kinds of farm implements, mill machinery, for piles and wharves, railway ties, etc., etc. The oaks are medium- to large-sized trees, forming the predominant part of a large proportion of our broad-leaved [71] forests, so that these are generally termed "oak forests," though they always contain considerable proportion of other kinds of trees. Three well-marked kinds — white, red, and live oak — are distinguished and kept separate in the markets. Of the two principal kinds "white oak" is the stronger, tougher, less porous, and more durable. "Red oak" is usually of coarser texture, more porous, often brittle, less durable, and even more troublesome in seasoning than white oak. In carpentry and furniture work red oak brings the same price at present as white oak. The red oaks everywhere accompany the white oaks, and, like the latter, are usually represented by several species in any given locality. "Live oak," once largely employed in shipbuilding, possesses all the good qualities, except that of size, of white oak, even to a greater degree. It is one of the heaviest, hardest, toughest, and most durable woods of this country. In structure it resembles the red oak, but is less porous.

65. White Oak (*Quercus alba*) (American Oak). Medium- to large-sized tree. Heartwood light brown, sapwood lighter color. Annual rings well marked, medullary rays broad and prominent. Wood tough, strong, heavy, hard, liable to check in seasoning, durable in contact with the soil, takes a high polish, very elastic, does not shrink much, and can be bent to any form when steamed. Used for agricultural implements, tool handles, furniture, fixtures, interior finish, car and wagon construction, beams, cabinet work, tight cooperage, railway ties, etc., etc. Because of the broad medullary rays, it is generally "quarter-sawn" for cabinet work and furniture. Common in the Eastern States, Ohio and Mississippi Valleys. Occurs throughout the eastern United States.

66. White Oak (*Quercus durandii*). Medium- to small-sized tree. Wood in its quality and uses similiar to the preceding. Texas, eastward to Alabama.

67. White Oak (*Quercus garryana*) (Western White Oak). Medium- to large-sized tree. Stronger, more durable, [72] and wood more compact than *Quercus alba*. Washington to California.

68. White Oak (*Quercus lobata*). Medium- to large-sized tree. Largest oak on the Pacific Coast. Wood in its quality and uses similar to *Quercus alba*, only it is finer-grained. California.

69. Bur Oak (*Quercus macrocarpa*) (Mossy-Cup Oak, Over-Cup Oak). Large-sized tree. Heartwood "oak" brown, sapwood lighter color. Wood heavy, strong, close-grained, durable in contact with the soil. Used in ship- and boatbuilding, all sorts of construction, interior finish of houses, cabinet work, tight cooperage, carriage and wagon work, agricultural implements, railway ties, etc., etc. One of the most valuable and most widely distributed of American oaks, 60 to 80 feet in height, and, unlike most of the other oaks, adapts itself to varying climatic conditions. It is one of the most durable woods when in contact with the soil. Common, locally abundant. Ranges from Manitoba to Texas, and from the foot hills of the Rocky Mountains to the Atlantic Coast. It is the most abundant oak of Kansas and Nebraska, and forms the scattered forests known as "The oak openings" of Minnesota.

70. Willow Oak (*Quercus phellos*) (Peach oak). Small to medium-sized tree. Heartwood pale reddish brown, sapwood lighter color. Wood heavy, hard, strong, coarse-grained. Occasionally used in construction. New York to Texas, and northward to Kentucky.

71. Swamp White Oak (*Quercus bicolor* var. *platanoides*). Large-sized tree. Heartwood pale brown, sapwood the same color. Wood heavy, hard, strong, tough, coarse-grained, checks considerably in seasoning. Used in construction, interior finish of houses, carriage- and boatbuilding, agricultural implements, in cooperage, railway ties, fencing, etc., etc. Ranges from Quebec to Georgia and westward to Arkansas. Never abundant. Most abundant in the Lake States. [73]

72. Over-Cup Oak (*Quercus lyrata*) (Swamp White Oak, Swamp Post Oak). Medium to large-sized tree, rather restricted, as it grows in the swampy districts of Carolina and Georgia. Is a larger tree than most of the other oaks, and produces an excellent timber, but grows in districts difficult of access, and is not much used. Lower Mississippi and eastward to Delaware.

73. Pin Oak (*Quercus palustris*) (Swamp Spanish Oak, Water Oak). Medium- to large-sized tree. Heartwood pale brown with dark-colored sap wood. Wood heavy, strong, and coarse-grained. Common along the borders of streams and swamps, attains its greatest size in the valley of the Ohio. Arkansas to Wisconsin, and eastward to the Alleghanies.

74. Water Oak (*Quercus aquatica*) (Duck Oak, Possum Oak). Medium- to large-sized tree, of extremely rapid growth. Eastern Gulf States, eastward to Delaware and northward to Missouri and Kentucky.

75. Chestnut Oak (*Quercus prinus*) (Yellow Oak, Rock Oak, Rock Chestnut Oak). Heartwood dark brown, sapwood lighter color. Wood heavy, hard, strong, tough, close-grained, durable in contact with the soil. Used for railway ties, fencing, fuel, and locally for construction. Ranges from Maine to Georgia and Alabama, westward through Ohio, and southward to Kentucky and Tennessee.

76. Yellow Oak (*Quercus acuminata*) (Chestnut Oak, Chinquapin Oak). Medium- to large-sized tree. Heartwood dark brown, sapwood pale brown. Wood heavy, hard, strong, close-grained, durable in contact with the soil. Used in the manufacture of wheel stock, in cooperage, for railway ties, fencing, etc., etc. Ranges from New York to Nebraska and eastern Kansas, southward in the Atlantic region to the District of Columbia, and west of the Alleghanies southward to the Gulf States. [74]

77. Chinquapin Oak (*Quercus prinoides*) (Dwarf Chinquapin Oak, Scrub Chestnut Oak). Small-sized tree. Heartwood light brown, sapwood darker color. Does not enter the markets to any great extent. Ranges from Massachusetts to North Carolina, westward to

Missouri, Nebraska, Kansas, and eastern Texas. Reaches its best form in Missouri and Kansas.

78. Basket Oak (*Quercus michauxii*) (Cow Oak). Large-sized tree. Locally abundant. Lower Mississippi and eastward to Delaware.

79. Scrub Oak (*Quercus ilicifolia* var. *pumila*) (Bear Oak). Small-sized tree. Heartwood light brown, sapwood darker color. Wood heavy, hard, strong, and coarse-grained. Found in New England and along the Alleghanies.

80. Post Oak (*Quercus obtusiloda* var. *minor*) (Iron Oak). Medium- to large-sized tree, gives timber of great strength. The color is of a brownish yellow hue, close-grained, and often superior to the white oak (*Quercus alba*) in strength and durability. It is used for posts and fencing, and locally for construction. Arkansas to Texas, eastward to New England and northward to Michigan.

81. Red Oak (*Quercus rubra*) (Black Oak). Medium- to large-sized tree. Heartwood light brown to red, sapwood lighter color. Wood coarse-grained, well-marked annual rings, medullary rays few but broad. Wood heavy, hard, strong, liable to check in seasoning. It is found over the same range as white oak, and is more plentiful. Wood is spongy in grain, moderately durable, but unfit for work requiring strength. Used for agricultural implements, furniture, bob sleds, vehicle parts, boxes, cooperage, woodenware, fixtures, interior finish, railway ties, etc., etc. Common in all parts of its range. Maine to Minnesota, and southward to the Gulf.

82. Black Oak (*Quercus tinctoria* var. *velutina*) (Yellow Oak). Medium- to large-sized tree. Heartwood [75] bright brown tinged with red, sapwood lighter color. Wood heavy, hard, strong, coarse-grained, checks considerably in seasoning. Very common in the Southern States, but occurring North as far as Minnesota, and eastward to Maine.

83. Barren Oak (*Quercus nigra* var. *marilandica*) (Black Jack, Jack Oak). Small-sized tree. Heartwood dark brown, sapwood lighter color. Wood heavy, hard, strong, coarse-grained, not valuable. Used in the manufacture of charcoal and for fuel. New York to Kansas and Nebraska, and southward to Florida. Rare in the North, but abundant in the South.

84. Shingle Oak (*Quercus imbricaria*) (Laurel Oak). Small to medium-sized tree. Heartwood pale reddish brown, sapwood lighter color. Wood heavy, hard, strong, coarse-grained, checks considerably in drying. Used for shingles and locally for construction. Rare in the east, most abundant in the lower Ohio Valley. From New York to Illinois and southward. Reaches its greatest size in southern Illinois and Indiana.

85. Spanish Oak (*Quercus digitata* var. *falcata*) (Red Oak). Medium-sized tree. Heartwood light reddish brown, sapwood much lighter. Wood heavy, hard, strong, coarse-grained, and checks considerably in seasoning. Used locally for construction, and has high fuel value. Common in south Atlantic and Gulf region, but found from Texas to New York, and northward to Missouri and Kentucky.

86. Scarlet Oak (*Quercus coccinea*). Medium- to large-sized tree. Heartwood light reddish-brown, sapwood darker color. Wood heavy, hard, strong, and coarse-grained. Best developed in the lower basin of the Ohio, but found from Minnesota to Florida.

87. Live Oak (*Quercus virens*) (Maul Oak). Medium- to large-sized tree. Grows from Maryland to the Gulf [76] of Mexico, and often attains a height of 60 feet and 4 feet in diameter. The wood is hard, strong, and durable, but of rather rapid growth, therefore not as good quality as *Quercus alba*. The live oak of Florida is now reserved by the United States Government for Naval purposes. Used for mauls and mallets, tool handles, etc., and locally for construction. Scattered along the coast from Maryland to Texas.

88. Live Oak (*Quercus chrysolepis*) (Maul Oak, Valparaiso Oak). Medium- to small-sized tree. California.

OSAGE ORANGE

89. Osage Orange (*Maclura aurantiaca*) (Bois d'Arc). A small-sized tree of fairly rapid growth. Wood very heavy, exceedingly hard, strong, not tough, of moderately coarse texture, and very durable and elastic. Sapwood yellow, heartwood brown on the end face, yellow on the longitudinal faces, soon turning grayish brown if exposed. It shrinks considerably in drying, but once dry it stands unusually well. Much used for wheel stock, and wagon framing; it is easily split, so is unfit for wheel hubs, but is very suitable for wheel spokes. It is considered one of the timbers likely to supply the place of black locust for insulator pins on telegraph poles. Seems too little appreciated; it is well suited for turned ware and especially for woodcarving. Used for spokes, insulator pins, posts, railway ties, wagon framing, turnery, and woodcarving. Scattered through the rich bottoms of Arkansas and Texas.

PAPAW

90. Papaw (*Asimina triloba*) (Custard Apple). Small-sized tree, often only a shrub, Heartwood pale, yellowish green, sapwood lighter color. Wood light, soft, coarse-grained, and spongy. Not used to any extent in manufacture. Occurs in eastern and central Pennsylvania, west as far as Michigan and Kansas, and south to Florida and Texas. Often forming [77] dense thickets in the lowlands bordering the Mississippi River.

PERSIMMON

91. Persimmon (*Diospyros Virginiana*). Small to medium-sized tree. Wood very heavy, and hard, strong and tough; resembles hickory, but is of finer texture and elastic, but liable to split in working. The broad sapwood cream color, the heartwood brown, sometimes almost black. The persimmon is the Virginia date plum, a tree of 30 to 50 feet high, and 18 to 20 inches in diameter; it is noted chiefly for its fruit, but it produces a wood of considerable value. Used in turnery, for wood engraving, shuttles, bobbins, plane stock, shoe lasts, and largely as a substitute for box (*Buxus sempervirens*)—especially the black or Mexican variety,—also used for pocket rules and drawing scales, for flutes and other wind instruments. Common, and best developed in the lower Ohio Valley, but occurs from New York to Texas and Missouri.

POPLAR (See also Tulip Wood)

Wood light, very soft, not strong, of fine texture, and whitish, grayish to yellowish color, usually with a satiny luster. The wood shrinks moderately (some cross-grained forms warp excessively), but checks very little in seasoning; is easily worked, but is not durable. Used in cooperage, for building and furniture lumber, for crates and boxes (especially cracker boxes), for woodenware, and paper pulp.

92. Cottonwood (*Populus monilifera*, var. *angulata*) (Carolina Poplar). Large-sized tree, forms considerable forests along many of the Western streams, and furnishes most of the cottonwood of the market. Heartwood dark brown, sapwood nearly white. Wood light, soft, not strong, and close-grained (see Fig. 14). Mississippi Valley and West. New England to the Rocky Mountains. [78]

93. Cottonwood (*Populus fremontii* var. *wislizeni*). Medium- to large-sized tree. Common. Wood in its quality and uses similiar to the preceding, but not so valuable. Texas to California.

Fig. 14. A Large Cottonwood. One of the Associates of Red Gum.

94. Black Cottonwood (*Populus trichocarpa* var. *heterophylla*) (Swamp Cottonwood, Downy Poplar). The largest deciduous tree of Washington. Very common. [79] Heartwood dull brown, sapwood lighter brown. Wood soft, close-grained. Is now manufactured into lumber in the West and South, and used in interior finish of buildings. Northern Rocky Mountains and Pacific region.

95. Poplar (*Populus grandidentata*) (Large-Toothed Aspen). Medium-sized tree. Heartwood light brown, sapwood nearly white. Wood soft and close-grained, neither strong nor durable. Chiefly used for wood pulp. Maine to Minnesota and southward along the Alleghanies.

96. White Poplar (*Populus alba*) (Abele-Tree). Small to medium-sized tree. Wood in its quality and uses similar to the preceding. Found principally along banks of streams, never forming forests. Widely distributed in the United States.

97. Lombardy Poplar (*Populus nigra italica*). Medium- to large-sized tree. This species is the first ornamental tree introduced into the United States, and originated in Afghanistan. Does not enter into the markets. Widely planted in the United States.

98. Balsam (*Populus balsamifera*) (Balm of Gilead, Tacmahac). Medium- to large-sized tree. Heartwood light brown, sapwood nearly white. Wood light, soft, not strong, close-grained. Used extensively in the manufacture of paper pulp. Common all along the northern boundary of the United States.

99. Aspen (*Populus tremuloides*) (Quaking Aspen). Small to medium-sized tree, often forming extensive forests, and covering burned areas. Heartwood light brown, sapwood nearly white. Wood light, soft, close-grained, neither strong nor durable. Chiefly used for woodenware, cooperage, and paper pulp. Maine to Washington and northward, and south in the western mountains to California and New Mexico.

RED GUM (See Gum) [80]

SASSAFRAS

100. Sassafras (*Sassafras sassafras*). Medium-sized tree, largest in the lower Mississippi Valley. Wood light, soft, not strong, brittle, of coarse texture, durable in contact with the soil. The sapwood yellow, the heartwood orange brown. Used to some extent in slack cooperage, for skiff- and boatbuilding, fencing, posts, sills, etc. Occurs from New England to Texas and from Michigan to Florida.

SOUR GUM (See Gum)

SOURWOOD

101. Sourwood (*Oxydendrum arboreum*) (Sorrel-Tree). A slender tree, reaching the maximum height of 60 feet. Heartwood reddish brown, sapwood lighter color. Wood heavy, hard, strong, close-grained, and takes a fine polish. Ranges from Pennsylvania, along the Alleghanies, to Florida and Alabama, westward through Ohio to southern Indiana and southward through Arkansas and Louisiana to the Coast.

SWEET GUM (See Gum)

SYCAMORE

102. Sycamore (*Platanus occidentalis*) (Buttonwood, Button-Ball Tree, Plane Tree, Water Beech). A large-sized tree, of rapid growth. One of the largest deciduous trees of the United States, sometimes attaining a height of 100 feet. It produces a timber that is moderately heavy, quite hard, stiff, strong, and tough, usually cross-grained; of coarse texture, difficult to split and work, shrinks moderately, but warps and checks considerably in seasoning, but stands well, and is not considered durable for outside work, or in contact with the soil. It has broad medullary rays, and much of the timber has a beautiful figure. It is used in slack cooperage, and quite extensively for [81] drawers, backs, and bottoms, etc., in furniture work. It is also used for cabinet work, for tobacco boxes, crates, desks, flooring, furniture, ox-yokes, butcher blocks, and also for finishing lumber, where it has too long been underrated. Common and largest in the Ohio and Mississippi Valleys, at home in nearly all parts of the eastern United States.

103. Sycamore (*Platanus racemosa*). The California species, resembling in its wood the Eastern form. Not used to any great extent.

TULIP TREE

104. Tulip Tree (*Liriodendron tulipifera*) (Yellow Poplar, Tulip Wood, White Wood, Canary Wood, Poplar, Blue Poplar, White Poplar, Hickory Poplar). A medium- to large-sized tree, does not form forests, but is quite common, especially in the Ohio basin. Wood usually light, but varies in weight, it is soft, tough, but not strong, of fine texture, and yellowish color. The wood shrinks considerably, but seasons without much injury, and works and stands extremely well. Heartwood light yellow or greenish brown, the sapwood is thin, nearly white, and decays rapidly. The heartwood is fairly durable when exposed to the weather or in contact with the soil. It bends readily when steamed, and takes stain and paint well. The mature forest-grown tree has a long, straight, cylindrical bole, clear of branches for at least two thirds of its length, surmounted by a short, open, irregular crown. When growing in the open, the tree maintains a straight stem, but the crown extends almost to the ground, and is of conical shape. Yellow poplar, or tulip wood, ordinarily grows to a height of from 100 to 125 feet, with a diameter of from 3 to 6 feet, and a clear length of about 70 feet. Trees have been found 190 feet high and ten feet in diameter. Used in cooperage, for siding, for panelling and finishing lumber in houses, car- and ship-building, for sideboards, panels of wagons and carriages, for aeroplanes, [82] for automobiles, also in the manufacture of furniture farm implements, machinery, for pump logs, and almost every kind of common woodenware, boxes shelving, drawers, etc., etc. Also in the manufacture of toys, culinary woodenware, and backing for veneer. It is in great demand throughout the vehicle and implement trade, and also makes a fair grade of wood pulp. In fact the tulip tree is one of the most useful of woods throughout the woodworking industry of this country. Occurs from New England to Missouri and southward to Florida.

TUPELO (See Gum)

WAAHOO

105. Waahoo (*Evonymus atropurpureus*). (Burning Bush, Spindle Tree). A small-sized tree. Wood white, tinged with orange; heavy, hard, tough, and close-grained, works well and stands well. Used principally for arrows and spindles. Widely distributed. Usually a shrub six to ten feet high, becoming a tree only in southern Arkansas and Oklahoma.

WALNUT

106. Black Walnut (*Juglans nigra*) (Walnut). A large, beautiful, and quickly-growing tree, about 60 feet and upwards in height. Wood heavy, hard, strong, of coarse texture, very durable in contact with the soil. The narrow sapwood whitish, the heartwood dark, rich, chocolate brown, sometimes almost black; aged trees of fine quality bring fancy prices. The wood shrinks moderately in seasoning, works well and stands well, and takes a fine polish. It is quite handsome, and has been for a long time the favorite wood for cabinet and furniture making. It is used for gun-stocks, fixtures, interior decoration, veneer, panelling, stair newells, and all classes of work demanding a high priced grade of wood. Black walnut is a large tree with stout trunk, of rapid growth, and [83] was formerly quite abundant throughout the Alleghany region. Occurs from New England to Texas, and from Michigan to Florida. Not common.

WHITE WALNUT (See Butternut)

WHITE WOOD (See Tulip and also Basswood)

WHITE WILLOW

107. White Willow (*Salix alba* var. *vitellina*) (Willow, Yellow Willow, Blue Willow). The wood is very soft, light, flexible, and fairly strong, is fairly durable in contact with the soil, works well and stands well when seasoned. Medium-sized tree, characterized by a short, thick trunk, and a large, rather irregular crown composed of many branches. The size of the tree at maturity varies with the locality. In the region where it occurs naturally, a height of 70 to 80 feet, and a diameter of three to four feet are often attained. When planted in the Middle West, a height of from 50 to 60 feet, and a diameter of one and one-half to two feet are all that may be expected. When closely planted on moist soil, the tree forms a tall, slender stem, well cleared branches. Is widely naturalized in the United States. It is used in cooperage, for woodenware, for cricket and baseball bats, for basket work, etc. Charcoal made from the wood is used in the manufacture of gunpowder. It has been generally used for fence posts on the Northwestern plains, because of scarcity of better material. Well seasoned posts will last from four to seven years. Widely distributed throughout the United States.

108. Black Willow (*Salix nigra*). Small-sized tree. Heartwood light reddish brown, sapwood nearly white. Wood soft, light, not strong, close-grained, and very flexible. Used in basket making, etc. Ranges from New York to Rocky Mountains and southward to Mexico. [84]

109. Shining Willow (*Salix lucida*). A small-sized tree. Wood in its quality and uses similiar to the preceding. Ranges from Newfoundland to Rocky Mountains and southward to Pennsylvania and Nebraska.

110. Perch Willow (*Salix amygdaloides*) (Almond-leaf Willow). Small to medium-sized tree. Heartwood light brown, sapwood lighter color. Wood light, soft, flexible, not strong, close-grained. Uses similiar to the preceding. Follows the water courses and ranges across the continent; less abundant in New England than elsewhere. Common in the West.

111. Long-Leaf Willow (*Salix fluviatilis*) (Sand Bar Willow). A small-sized tree. Ranges from the Arctic Circle to Northern Mexico.

112. Bebb Willow (*Salix bebbiana* var. *rostrata*). A small-sized tree. More abundant in British America than in the United States, where it ranges southward to Pennsylvania and westward to Minnesota.

113. Glaucous Willow (*Salix discolor*) (Pussy Willow). A small-sized tree. Common along the banks of streams, and ranges from Nova Scotia to Manitoba, and south to Delaware; west to Indiana and northwestern Missouri.

114. Crack Willow (*Salix fragilis*). A medium to large-sized tree. Wood is very soft, light, very flexible and fairly strong, is fairly durable in contact with the soil, works well and stands well. Used principally for basket making, hoops, etc., and to produce charcoal for gunpowder. Very common, and widely distributed in the United States.

115. Weeping Willow (*Salix babylonica*). Medium- to large-sized tree. Wood similiar to *Salix nigra*, but not so valuable. Mostly an ornamental tree. Originally came from China. Widely planted in the United States. [85]

YELLOW WOOD

116. Yellow Wood (*Cladrastis lutea*) (Virgilia). A small to medium-sized tree. Wood yellow to pale brown, heavy, hard, close-grained and strong. Not used to much extent in manufacturing. Not common. Found principally on the limestone cliffs of Kentucky, Tennessee, and North Carolina.

GRAIN, COLOR, ODOR, WEIGHT, AND FIGURE IN WOOD

DIFFERENT GRAINS OF WOOD

The terms "fine-grained," "coarse-grained," "straight-grained," and "cross-grained" are frequently applied in the trade. In common usage, wood is coarse-grained if its annual rings are wide; fine-grained if they are narrow. In the finer wood industries a fine-grained wood is capable of high polish, while a coarse-grained wood is not, so that in this latter case the distinction depends chiefly on hardness, and in the former on an accidental case of slow or rapid growth. Generally if the direction of the wood fibres is parallel to the axis of the stem or limb in which they occur, the wood is straight-grained; but in many cases the course of the fibres is spiral or twisted around the tree (as shown in Fig. 15), and sometimes commonly in the butts of gum and cypress, the fibres of several layers are oblique in one direction, and those of the next series of layers are oblique in the opposite direction. (As shown in Fig. 16 the wood is cross or twisted grain.) Wavy-grain in a tangential plane as seen on the radial section is illustrated in Fig. 17, which represents an extreme case observed in beech. This same form also occurs on the radial plane, causing the tangential section to appear wavy or in transverse folds.

When wavy grain is fine (*i.e.*, the folds or ridges small but numerous) it gives rise to the "curly" structure frequently seen in maple. Ordinarily, neither wavy, spiral, nor alternate grain is visible on the cross-section; its existence often escapes the eye even on smooth, longitudinal faces in the sawed material, so that the only [87] guide to their discovery lies in splitting the wood in two, in the two normal plains.

Fig. 15. Spiral Grain. Season checks, after removal of bark, indicate the direction of the fibres or grain of the wood.

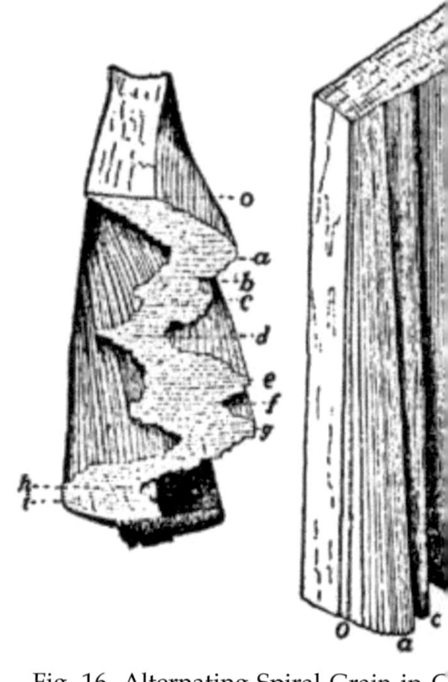

Fig. 16. Alternating Spiral Grain in C end view of same piece. When the ba grain of this piece was straight. From th it grew more oblique in one direction, r at *a*, and then turned back in the op These alternations were repeated peric sharing in these changes.

Generally the surface of the wood under the bark, and therefore also that of any layer in the interior, is not uniform and smooth, but is channelled and pitted by numerous depressions, which differ greatly in size and form. Usually, any one depression or elevation is restricted to one or few annual layers (*i.e.*, seen only in one or few rings) and is then lost, being compensated (the surface at the particular spot evened up) by growth. In some woods, however, any depression or elevation once attained grows from year to year and

100

reaches a maximum size, which is maintained for many years, sometimes throughout life. In maple, where this tendency to preserve any particular contour is very great, the depressions and elevations are [88] usually small (commonly less than one-eighth inch) but very numerous.

On tangent boards of such wood, the sections, pits, and prominences appear as circlets, and give rise to the beautiful "bird's eye" or "landscape" structure. Similiar structures in the burls of black ash, maple, etc., are frequently due to the presence of dormant buds, which cause the surface of all the layers through which they pass to be covered by small conical elevations, whose cross-sections on the sawed board appear as irregular circlets or islets, each with a dark speck, the section of the pith or "trace" of the dormant bud in the center.

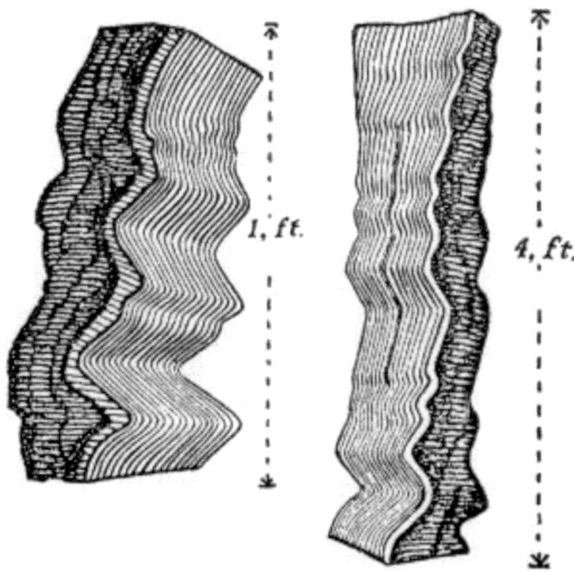

Fig. 17. Wavy Grain in Beech (*after Nordlinger*).

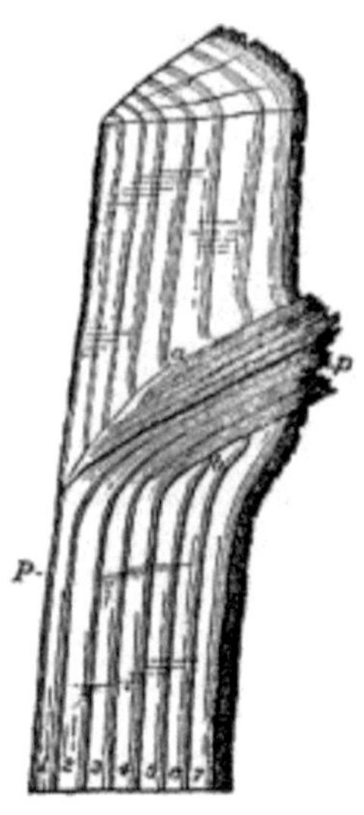

In the wood of many broad-leaved trees the wood fibres are much longer when full grown than when they are first formed in the cambium or growing zone. This causes the tips of each fibre to crowd in between the fibres above and below, and leads to an irregular interlacement of these fibres, which adds to the toughness, but reduces the cleavability of the wood. At the juncture of the limb and stem the fibres on the upper and lower sides of the limb behave [89] differently. On the lower side they run from the stem into the limb, forming an uninterrupted strand or tissue and a perfect union. On the upper side the fibres bend aside, are not continuous into the limb, and hence the connection is not perfect (see Fig. 18). Owing to this arrangement of the fibres, the cleft made in splitting never runs into the knot if started on the side above the limb, but is apt to enter the knot if started below, a fact well understood in woodcraft. When limbs die, decay, and break off, the remaining stubs are surrounded, and may finally be covered by the growth of the trunk and thus give rise to the annoying "dead" or "loose" knots.

> Fig. 18. Section of Wood showing Position of the Grain at Base of a Limb. P, pith of both stem and limb; 1-7, seven yearly layers of wood; *a, b,* knot or basal part of a limb which lived for four years, then died and broke off near the stem, leaving the part to the left of *a, b,* a

"sound" knot, the part to the right a "dead" knot, which would soon be entirely covered by the growing stem.

COLOR AND ODOR OF WOOD

Color, like structure, lends beauty to the wood, aids in its identification, and is of great value in the determination of its quality. If we consider only the heartwood, the black color of the persimmon, the dark brown of the walnut, the light brown of the white oaks, the reddish brown of the red oaks, the yellowish white of the tulip and poplars, the brownish red of the redwood and cedars, the yellow of the papaw and sumac, are all reliable marks of distinction and color. Together with luster and weight, they are only too often the only features depended upon in practice. Newly formed wood, like that of the outer few rings, has but little color. The sapwood generally is light, [90] and the wood of trees which form no heartwood changes but little, except when stained by forerunners of disease.

The different tints of colors, whether the brown of oak, the orange brown of pine, the blackish tint of walnut, or the reddish cast of cedar, are due to pigments, while the deeper shade of the summerwood bands in pine, cedar, oak, or walnut is due to the fact that the wood being denser, more of the colored wood substance occurs on a given space, *i.e.*, there is more colored matter per square inch. Wood is translucent, a thin disk of pine permitting light to pass through quite freely. This translucency affects the luster and brightness of lumber.

When lumber is attacked by fungi, it becomes more opaque, loses its brightness, and in practice is designated "dead," in distinction to "live" or bright timber. Exposure to air darkens all wood; direct sunlight and occasional moistening hasten this change, and cause it to penetrate deeper. Prolonged immersion has the same effect, pine wood becoming a dark gray, while oak changes to a blackish brown.

Odor, like color, depends on chemical compounds, forming no part of the wood substance itself. Exposure to weather reduces and often changes the odor, but a piece of long-leaf pine, cedar, or camphor wood exhales apparently as much odor as ever when a new surface is exposed. Heartwood is more odoriferous than sapwood.

Many kinds of wood are distinguished by strong and peculiar odors. This is especially the case with camphor, cedar, pine, oak, and mahogany, and the list would comprise every kind of wood in use were our sense of smell developed in keeping with its importance.

Decomposition is usually accompanied by pronounced odors. Decaying poplar emits a disagreeable odor, while red oak often becomes fragrant, its smell resembling that of heliotrope. [91]

WEIGHT OF WOOD

A small cross-section of wood (as in Fig. 19) dropped into water sinks, showing that the substance of which wood fibre or wood is built up is heavier than water. By immersing the wood successively in heavier liquids, until we find a liquid in which it does not sink, and comparing the weight of the same with water, we find that wood substance is about 1.6 times as heavy as water, and that this is as true of poplar as of oak or pine.

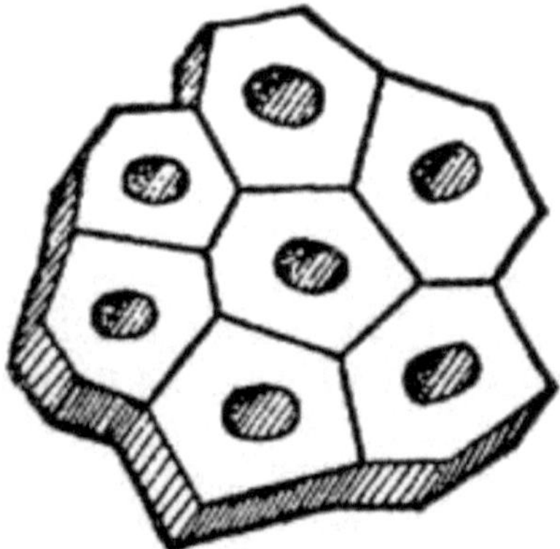

Fig. 19. Cross-section of a Group of Wood Fibres (Highly Magnified.)

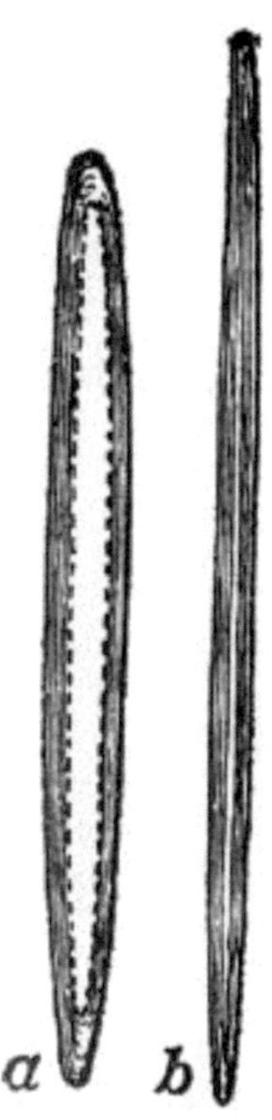

Fig. 20. Isolated Fibres of Wood.

Separating a single cell (as shown in Fig. 20, *a*), drying and then dropping it into water, it floats. The air-filled cell cavity or interior reduces its weight, and, like an empty corked bottle, it weighs less than the water. Soon, however, water soaks into the cell, when it fills up and sinks. Many such cells grown together, as in a block of wood, when all or most of them are filled with water, will float as long as the majority of them are empty or only partially filled. This is why a green, sappy pine pole soon sinks in "driving" (floating down stream). Its cells are largely filled before it is thrown in, and but little additional water suffices to make its weight greater than that of the water. In a good-sized white pine log, composed chiefly of empty cells (heartwood), the water requires a very long time to fill up the cells (five years would not suffice to fill them all), and therefore the log may float for many months. When the wall of the wood fibre is very thick (five eighths or more of the volume, as in Fig. 20, *b*), the fibre sinks whether empty or filled. This applies to most of the fibres of the dark summer-wood bands in pines, and to the compact fibres of oak or hickory, and many, especially tropical woods, [92] have such thick-walled cells and so little empty or air space that they never float.

Here, then, are the two main factors of weight in wood; the amount of cell wall or wood substance constant for any given piece, and the amount of water contained in the wood, variable even in the standing tree, and only in part eliminated in drying.

The weight of the green wood of any species varies chiefly as a second factor, and is entirely misleading, if the relative weight of different kinds is sought. Thus some green sticks of the otherwise lighter cypress and gum sink more readily than fresh oak.

The weight of sapwood or the sappy, peripheral part of our common lumber woods is always great, whether cut in winter or summer. It rarely falls much below forty-five pounds, and commonly exceeds fifty-five pounds to the cubic foot, even in our lighter wooded species. It follows that the green wood of a sapling is heavier than that of an old tree, the fresh wood from a disk of the upper part of a tree is often heavier than that of the lower part, and the wood near the bark heavier than that nearer the pith; and also that the advantage of drying the wood before shipping is most important in sappy and light kinds.

When kiln-dried, the misleading moisture factor of weight is uniformly reduced, and a fair comparison possible. For the sake of convenience in comparison, the weight of wood is expressed either as the weight per cubic foot, or, what is still more convenient, as specific weight or density. If an old long-leaf pine is cut up (as shown in Fig. 21) the wood of disk No. 1 is heavier than that of disk No. 2, the latter heavier than that of disk No. 3, and the wood of the top disk is found to be only about three fourths as heavy as that of disk No. 1. Similiarly, if disk No. 2 is cut up, as in the figure, the specific weight of the different parts is:

- *a*, about 0.52

- b, about 0.64
- c, about 0.67
- $d, e, f,$ about 0.65

showing that in this disk at least the wood formed during [93] the many years' growth, represented in piece a, is much lighter than that of former years. It also shows that the best wood is the middle part, with its large proportion of dark summer bands.

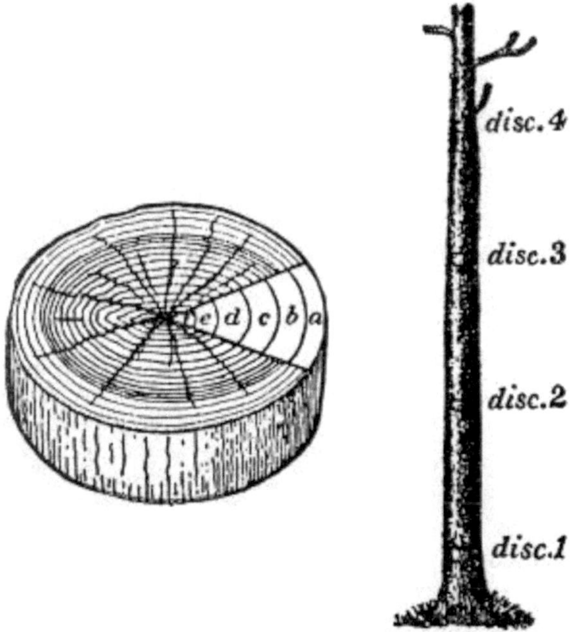

Fig. 21. Orientation of Wood Samples.

Cutting up all disks in the same way, it will be found that the piece a of the first disk is heavier than the piece a of the fifth, and that piece c of the first disk excels the piece c of all the other disks. This shows that the wood grown during the same number of years is lighter in the upper parts of the stem; and if the disks are smoothed on the radial surfaces and set up one on top of the other in their regular order, for the sake of comparison, this decrease in weight will be seen to be accompanied by a decrease in the amount of summer-wood. The color effect of the upper disks is conspicuously lighter. If our old pine had been cut one hundred and fifty years ago, before the outer, lighter wood was laid on, it is evident that the weight of the wood of any one disk would have been found to increase from the center outward, and no subsequent decrease could have been observed. [94]

In a thrifty young pine, then, the wood is heavier from the center outward, and lighter from below upward; only the wood laid on in old age falls in weight below the average. The number of brownish bands of summer-wood are a direct indication of these differences. If an old oak is cut up in the same manner, the butt cut is also found heaviest and the top lightest, but, unlike the disk of pine, the disk of oak has its firmest wood at the center, and each successive piece from the center outward is lighter than its neighbor.

Examining the pieces, this difference is not as readily explained by the appearance of each piece as in the case of pine wood. Nevertheless, one conspicuous point appears at once. The pores, so very distinct in oak, are very minute in the wood near the center, and thus the wood is far less porous.

Studying different trees, it is found that in the pines, wood with narrow rings is just as heavy as and often heavier than the wood with wider rings; but if the rings are unusually narrow in any part of the disk, the wood has a lighter color; that is, there is less summer-wood and therefore less weight.

In oak, ash, or elm trees of thrifty growth, the rings, fairly wide (not less than one-twelfth inch), always form the heaviest wood, while any piece with very narrow rings is light. On the other hand, the weight of a piece of hard maple or birch is quite independent of the width of its rings.

The bases of limbs (knots) are usually heavy, very heavy in conifers, and also the wood which surrounds them, but generally the wood of the limbs is lighter than that of the stem, and the wood of the roots is the lightest.

In general, it may be said that none of the native woods in common use in this country are when dry as heavy as water, *i.e.*, sixty-two pounds to the cubic foot. Few exceed fifty pounds, while most of them fall below forty pounds, and much of the pine and other coniferous wood weigh less than thirty pounds per cubic foot. The weight of the wood is in itself an important quality. Weight [95] assists in distinguishing maple from poplar. Lightness coupled with great strength and stiffness recommends wood for a thousand different uses. To a large extent weight predicates the strength of the wood, at least in the same species, so that a heavy piece of oak will exceed in strength a light piece of the same species, and in pine it appears probable that, weight for weight, the strength of the wood of various pines is nearly equal.

Weight of Kiln-dried Wood of Different Species

Species	Specific Weight	Approximate Weight of 1 Cubic Foot	1,000 Feet Lumber
(*a*) Very Heavy Woods: Hickory, Oak, Persimmon, Osage Orange, Black Locust, Hackberry, Blue Beech, best of Elm and Ash	0.70-0.80	42-48	3,700
(*b*) Heavy Woods Ash, Elm, Cherry, Birch, Maple, Beech, Walnut, Sour Gum, Coffee Tree, Honey Locust, best of Southern Pine and Tamarack	0.60-0.70	36-42	3,200
(*c*) Woods of Medium Weight: South-	0.50-	30-36	2,700

ern Pine, Pitch Pine, Tamarack, Douglas Spruce, Western Hemlock, Sweet Gum, Soft Maple, Sycamore, Sassafras, Mulberry, light grades of Birch and Cherry	0.60		
(*d*) Light Woods: Norway and Bull Pine, Red Cedar, Cypress, Hemlock, the Heavier Spruces and Firs, Redwood, Basswood, Chestnut, Butternut, Tulip, Catalpa, Buckeye, heavier grades of Poplar	0.40-0.50	24-30	2,200
(*e*) Very Light Woods: White Pine, Spruce, Fir, White Cedar, Poplar	0.30-0.40	18-24	1,800

"FIGURE" IN WOOD [96]

Many theories have been propounded as to the cause of "figure" in timber; while it is true that all timber possesses "figure" in some degree, which is more noticeable if it be cut in certain ways, yet there are some woods in which it is more conspicuous than in others, and which for cabinet or furniture work are much appreciated, as it adds to the value of the work produced.

The characteristic "figure" of oak is due to the broad and deep medullary rays so conspicuous in this timber, and the same applies to honeysuckle. Figure due to the same cause is found in sycamore and beech, but is not so pronounced. The beautiful figure in "bird's eye maple" is supposed to be due to the boring action of insects in the early growth of the tree, causing pits or grooves, which in time become filled up by being overlain by fresh layers of wood growth; these peculiar and unique markings are found only in the older and inner portion of the tree.

Pitch pine has sometimes a very beautiful "figure," but it generally does not go deep into the timber; walnut has quite a variety of "figures," and so has the elm. It is in mahogany, however, that we find the greatest variety of "figure," and as this timber is only used for furniture and fancy work, a good "figure" greatly enhances its value, as firmly figured logs bring fancy prices.

Mahogany, unlike the oak, never draws its "figure" from its small and almost unnoticeable medullary rays, but from the twisted condition of its fibres; the natural growth of mahogany produces a straight wood; what is called "figured" is unnatural and exceptional, and thus adds to its value as an ornamental wood. These peculiarities are rarely found in the earlier portion of the tree that is near the center, being in this respect quite different from maple; they appear when the tree is more fully developed, and consist of bundles of woody fibres which, instead of being laid in straight lines, behave in an erratic manner and are deposited in a twisted form; sometimes it may be caused by the intersection of branches, or possibly by the crackling of the bark pressing on the wood, and thus [97] moving it out of its natural straight course, causing a wavy line which in time becomes accentuated.

It will have been observed by most people that the outer portion of a tree is often indented by the bark, and the outer rings often follow a sinuous course which corresponds to this indention, but in most trees, after a few years, this is evened up and the annual rings assume their nearly circular form; it is supposed by some that in the case of mahogany this is not the case, and that the indentations are even accentuated.

The best figured logs of timber are secured from trees which grow in firm rocky soil; those growing on low-lying or swampy ground are seldom figured. To the practical woodworker the figure in mahogany causes some difficulty in planing the wood to a smooth surface; some portions plane smooth, others are the "wrong way of the grain."

Figure in wood is effected by the way light is thrown upon it, showing light if seen from one direction, and dark if viewed from another, as may easily be observed by holding a piece of figured mahogany under artificial light and looking at it from opposite directions. The characteristic markings on mahogany are "mottle," which is also found in sycamore, and is conspicuous on the backs of fiddles and violins, and is not in itself valuable; it runs the transverse way of the fibres and is probably the effect of the wind upon the tree in its early stages of growth. "Roe," which is said to be caused by the contortion of the woody fibres, and takes a wavy line parallel to them, is also found in the hollow of bent stems and in the root structure, and when combined with "mottle" is very valuable. "Dapple" is an exaggerated form of mottle. "Thunder shake," "wind shake," or "tornado shake" is a rupture of the fibres across the grain, which in mahogany does not always break them; the tree swaying in the wind only strains its fibres, and thus produces mottle in the wood.

SECTION V [98]

ENEMIES OF WOOD

From the writer's personal investigations of this subject in different sections of the country, the damage to forest products of various kinds from this cause seems to be far more extensive than is generally recognized. Allowing a loss of five per cent on the total value of the forest products of the country, which the writer believes to be a conservative estimate, it would amount to something over $30,000,000 annually. This loss differs from that resulting from insect damage to natural forest resources, in that it represents more directly a loss of money invested in material and labor. In dealing with the insects mentioned, as with forest insects in general, the methods which yield the best results are those which relate directly to preventing attack, as well as those which are unattractive or unfavorable. The insects have two objects in their attack: one is to obtain food, the other is to prepare for the development of their broods. Different species of insects have special periods during the season of activity (March to November), when the adults are on the wing in search of suitable material in which to deposit their eggs. Some species, which fly in April, will be attracted to the trunks of recently felled pine trees or to piles of pine sawlogs from trees felled the previous winter. They are not attracted to any other kind of timber, because they can live only in the bark or wood of pine, and only in that which is in the proper condition to favor the hatching of their eggs and the normal development of their young. As they fly only in April, they cannot injure the logs of trees felled during the remainder of the year. [99]

There are also oak insects, which attack nothing but oak; hickory, cypress, and spruce insects, etc., which have different habits and different periods of flight, and require special conditions of the bark and wood for depositing their eggs or for subsequent development of their broods.

Some of these insects have but one generation in a year, others have two or more, while some require more than one year for the complete development and transformation. Some species deposit their eggs in the bark or wood of trees soon after they are felled or before any perceptible change from the normal living tissue has taken place; other species are attracted only to dead bark and dead wood of trees which have been felled or girdled for several months; others are attracted to dry and seasoned wood; while another class will attack nothing but very old, dry bark or wood of special kinds and under special conditions. Thus it will be seen how important it is for the practical man to have knowledge of such of the foregoing facts as apply to his immediate interest in the manufacture or utilization of a given forest product, in order that he may with the least trouble and expense adjust his business methods to meet the requirements for preventing losses.

The work of different kinds of insects, as represented by special injuries to forest products, is the first thing to attract attention, and the distinctive character of this work is easily observed, while the insect responsible for it is seldom seen, or it is so difficult to determine by the general observer from descriptions or illustrations that the species is rarely recognized. Fortunately, the character of the work is often sufficient in itself to identify the cause and suggest a remedy, and in this section primary consideration is given to this phase of the subject.

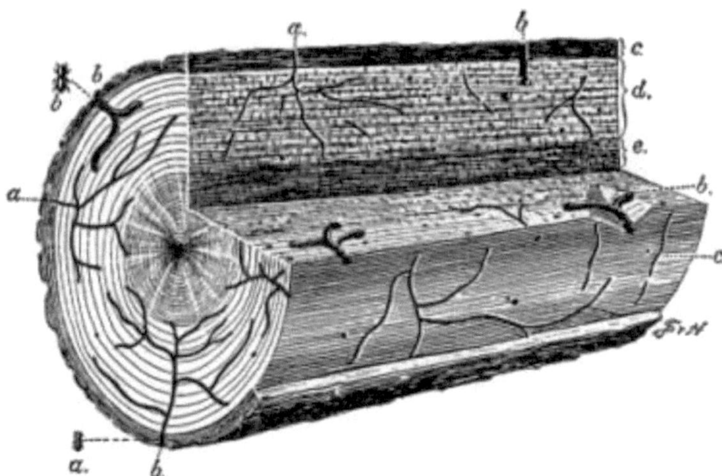

Fig. 22. Work of Ambrosia Beetles in Tulip or Yellow Poplar Wood. *a*, work of *Xyleborus affinis* and *Xyleborus inermis*; *b*, *Xyleborus obesus* and work; *c*, bark; *d*, sapwood; *e*, heartwood.

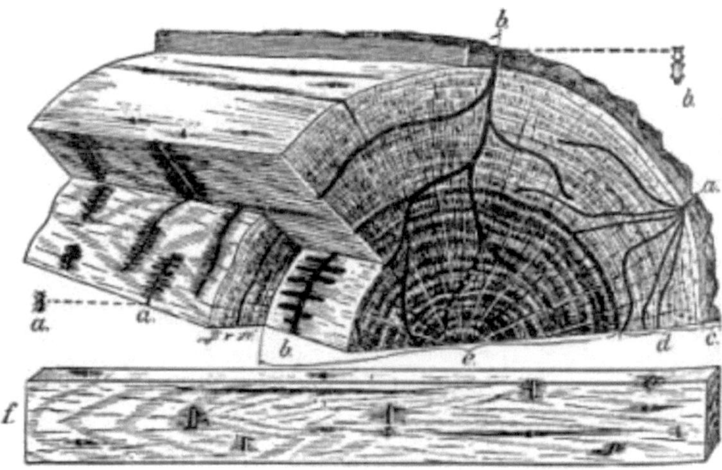

Fig. 23. Work of Ambrosia Beetles in Oak. *a*, *Monarthrum mali* and work; *b*, *Platypus compositus* and work; *c*, bark; *d*,

sapwood; *e*, heartwood; *f*, character of work in wood from injured log.

The characteristic work of this class of wood-boring beetles is shown in Figs. 22 and 23. The injury consists of pinhole and stained-wood defects in the sapwood and heartwood of recently felled or girdled trees, sawlogs, pulpwood, stave and shingle bolts, green or unseasoned [100] lumber, and staves and heads of barrels containing alcoholic liquids. The holes and galleries are made by the adult parent beetles, to serve as entrances and temporary houses or nurseries for the development of their broods of young, which feed on a fungus growing on the walls of the galleries.

The growth of this ambrosia-like fungus is induced and controlled by the parent beetles, and the young are [101] dependent upon it for food. The wood must be in exactly the proper condition for the growth of the fungus in order to attract the beetles and induce them to excavate their galleries; it must have a certain degree of moisture and other favorable qualities, which usually prevail during the period involved in the change from living, or normal, to dead or dry wood; such a condition is found in recently felled trees, sawlogs, or like crude products.

There are two general types or classes of these galleries: one in which the broods develop together in the main burrows (see Fig. 22), the other in which the individuals develop in short, separate side chambers, extending at right angles from the primary galleries (see Fig. 23). The galleries of the latter type are usually accompanied by a distinct staining of the wood, while those of the former are not.

The beetles responsible for this work are cylindrical in form, apparently with a head (the prothorax) half as long as the remainder of the body (see Figs. 22, *a*, and 23, *a*).

North American species vary in size from less than onetenth to slightly more than two-tenths of an inch, while some of the subtropical and tropical species attain a much larger size. The diameter of the holes made by each species corresponds closely to that of the body, and varies from about one-twentieth to one-sixteenth of an inch for the tropical species.

Round-headed Borers

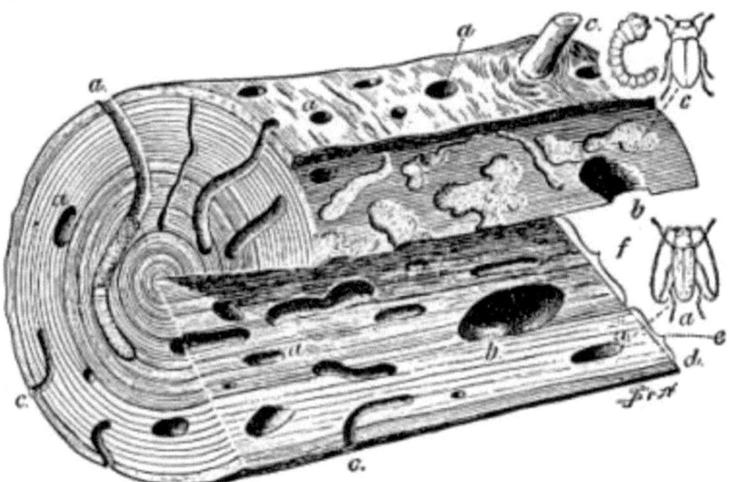

Fig. 24. Work of Round-headed and Flat-headed Borers in Pine. *a*, work of round-headed borer, "sawyer," *Monohammus spiculatus*, natural size *b*, *Ergates spiculatus*; *c*, work of flat-headed borer, *Buprestis*, larva and adult; *d*, bark; *e*, sapwood; *f*, heartwood.

The character of the work of this class of wood- and bark-boring grubs is shown in Fig. 24. The injuries consist of irregular flattened or nearly round wormhole defects in the wood, which sometimes result in the destruction [102] of valuable parts of the wood or bark material. The sapwood and heartwood of recently felled trees, sawlogs, poles posts, mine props, pulpwood and cordwood, also lumber or square timber, with bark on the edges, and construction timber in new and old buildings, are injured by wormhole defects, while the valuable parts of stored oak and hemlock tanbark and certain kinds of wood are converted into worm-dust. These injuries are caused by the young or larvae of long-horned beetles. Those which infest the wood hatch from eggs deposited in the outer bark of logs and like material, and the minute grubs hatching therefrom bore into the inner bark, through which they extend their irregular burrows, for the purpose of obtaining food from the sap and other nutritive material found in the plant tissue. They continue to extend and enlarge their burrows as they increase in size, until they are nearly or quite full grown. They then enter the wood and continue their excavations deep into the sapwood or heartwood until they attain their normal size. They then excavate pupa cells in which to transform into adults, [103] which emerge from the wood through exit holes in the surface. This class of borers is represented by a large number of species. The adults, however, are seldom seen by the general observer unless cut out of the wood before they have emerged.

Flat-headed Borers

The work of the flat-headed borers (Fig. 24) is only distinguished from that of the preceding by the broad, shallow burrows, and the much more oblong form of the exit holes. In general, the injuries are similar, and effect the same class of products, but they are of much less importance. The adult forms are flattened, metallic-colored beetles, and represent many species, of various sizes.

Timber Worms

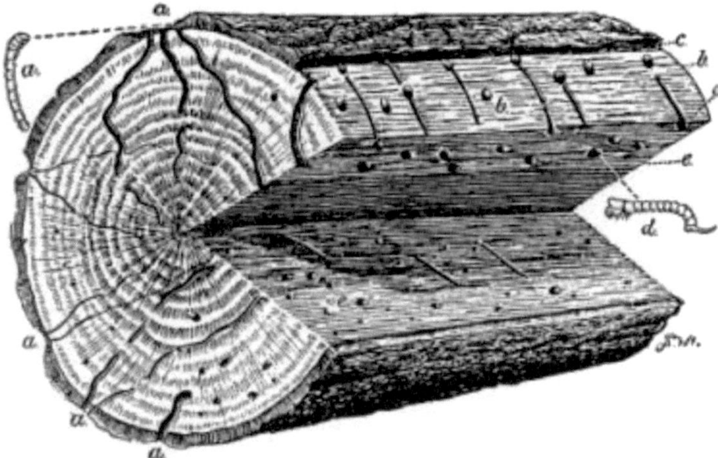

Fig. 25. Work of Timber Worms in Oak. *a*, work of oak timber worm, *Eupsalis minuta*; *b*, barked surface; *c*, bark; *d*, sapwood timber worm, *Hylocoetus lugubris*, and work; *e*, sapwood.

The character of the work done by this class is shown in Fig. 25. The injury consists of pinhole defects in the sapwood and heartwood of felled trees, sawlogs and like material which have been left in the woods or in piles in the open for several months during the warmer seasons. Stave [104] and shingle bolts and closely piled oak lumber and square timbers also suffer from injury of this kind. These injuries are made by elongate, slender worms or larvae, which hatch from eggs deposited by the adult beetles in the outer bark, or, where there is no bark, just beneath the surface of the wood. At first the young larvae bore almost invisible holes for a long distance through the sapwood and heartwood, but as they increase in size the same holes are enlarged and extended until the larvae have attained their full growth. They then transform to adults, and emerge through the enlarged entrance burrows. The work of these timber worms is distinguished from that of the timber beetles by the greater variation in the size of holes in the same piece of wood, also by the fact that they are not branched from a single entrance or gallery, as are those made by the beetles.

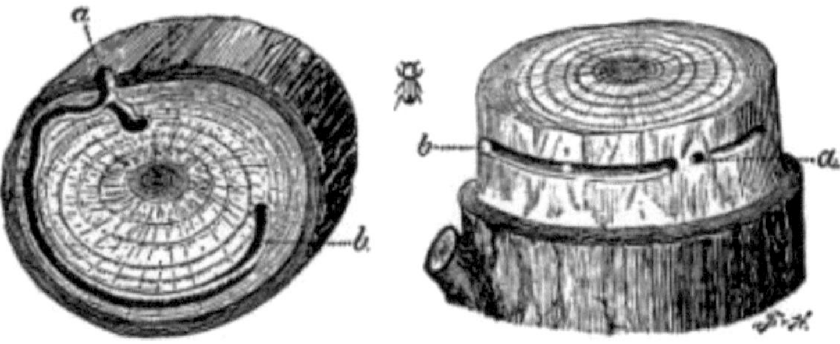

Fig. 26. Work of Powder Post Beetle, *Sinoxylon basilare*, in Hickory Poles, showing Transverse Egg Galleries excavated by the Adult, *a*, entrance; *b*, gallery; *c*, adult.

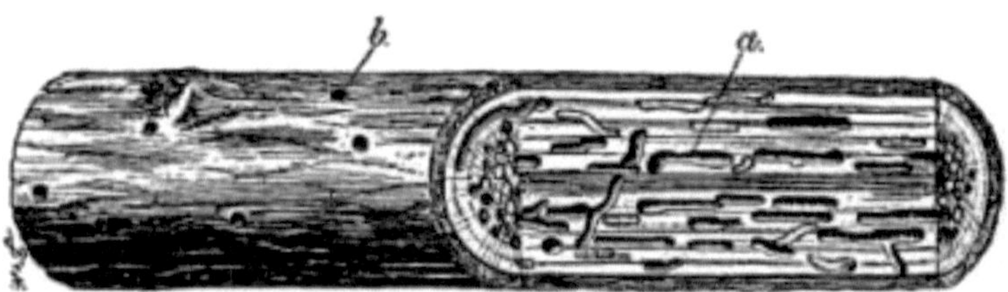

Fig. 27. Work of Powder Post Beetle, *Sinoxylon basilare*, in Hickory Pole. *a*, character of work by larvae; *b*, exit holes made by emerging broods. [105]

Powder Post Borers

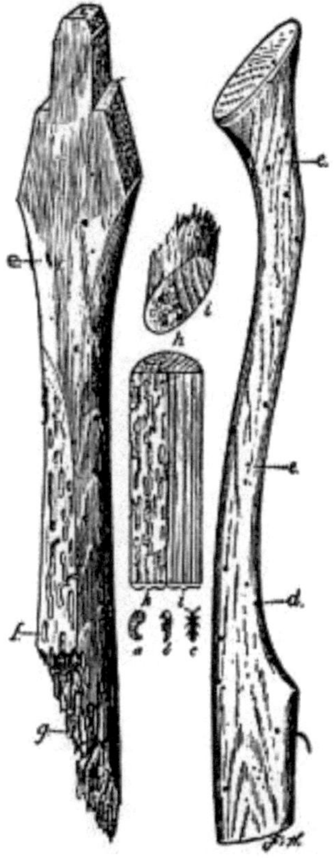

The character of the work of this class of insects is shown in Figs. 26, 27, and 28. The injury consists of closely placed burrows, packed with borings, or a completely destroyed or powdered condition of the wood of seasoned products, such as lumber, crude and finished handle and wagon stock, cooperage and wooden truss hoops, furniture, and inside finish woodwork, in old buildings, as well as in many other crude or finished and utilized woods. This is the work of both the adults and young stages of some species, or of the larval stage alone of others. In the former, the adult beetles deposit their eggs in burrows or galleries excavated for the purpose, as in Figs. 26 and 27, while in the latter (Fig. 28) the eggs are on or beneath the surface of the wood. The grubs complete the destruction by boring through the solid wood in all directions and packing their burrows with the powdered wood. When they are full grown they transform to the adult, and emerge from the injured material through holes in the surface. Some of the species continue to work in the same wood until many generations have developed and emerged or until every particle of wood tissue has been destroyed and the available nutritive substance extracted.

Fig. 28. Work of Powder Post Beetles, *Lyctus striatus*, in Hickory Handles and Spokes. *a*, larva; *b*, pupa; *c*, adult; *d*, exit holes; *e*, entrance of larvae (vents for borings are exits of parasites); *f*, work of larvae; *g*, wood, completely destroyed; *h*, sapwood; *i*, heartwood. [106]

Conditions Favorable for Insect Injury—Crude Products—Round Timber with Bark on

Newly felled trees, sawlogs, stave and heading bolts, telegraph poles, posts, and the like material, cut in the fall and winter, and left on the ground or in close piles during a few weeks or months in the spring or summer, causing them to heat and sweat, are especially liable to injury by ambrosia beetles (Figs. 22 and 23), round and flat-headed borers (Fig. 24), and timber worms (Fig. 25), as are also trees felled in the warm season, and left for a time before working up into lumber.

The proper degree of moisture found in freshly cut living or dying wood, and the period when the insects are flying, are the conditions most favorable for attack. This period of danger varies with the time of the year the timber is felled and with the different kinds of trees. Those felled in late fall and winter will generally remain attractive to ambrosia beetles, and to the adults of round- and flat-headed borers during March, April, and May. Those felled in April to September may be attacked in a few days after they are felled, and the period of danger may not extend over more than a few weeks. Certain kinds of trees felled during certain months and seasons are never attacked, because the danger period prevails only when the insects are flying; on the other hand, if the same kinds of trees are felled at a different time, the conditions may be most attractive when the insects are active, and they will be thickly infested and ruined.

The presence of bark is absolutely necessary for infestation by most of the wood-boring grubs, since the eggs and young stages must occupy the outer and inner portions before they can enter the wood. Some ambrosia and timber worms will, however, attack barked logs, especially those in close piles, and others shaded and protected from rapid drying.

The sapwood of pine, spruce, fir, cedar, cypress, and the like softwoods is especially liable to injury by ambrosia beetles, while the heartwood is sometimes ruined by a class of round-headed borers, known as "sawyers." Yellow [107] poplar, oak, chestnut, gum, hickory, and most other hardwoods are as a rule attacked by species of ambrosia beetles, sawyers, and timber worms, different from those infesting the pines, there being but very few species which attack both.

Mahogany and other rare and valuable woods imported from the tropics to this country in the form of round logs, with or without bark on, are commonly damaged more or less seriously by ambrosia beetles and timber worms.

It would appear from the writer's investigations of logs received at the mills in this country, that the principal damage is done during a limited period—from the time the trees are felled until they are placed in fresh or salt water for transportation to the shipping points. If, however, the logs are loaded on a vessel direct from the shore, or if not left in the water long enough to kill the insects, the latter will continue their destructive work during transportation to other countries and after they arrive, and until cold weather ensues or the logs are converted into lumber.

It was also found that a thorough soaking in sea-water, while it usually killed the insects at the time, did not prevent subsequent attacks by both foreign and native ambrosia beetles; also, that the removal of the bark from such logs previous to immersion did not render them entirely immune. Those with the bark off were attacked more than those with it on, owing to a greater amount of saline moisture retained by the bark.

How to Prevent Injury

From the foregoing it will be seen that some requisites for preventing these insect injuries to round timber are:

1. To provide for as little delay as possible between the felling of the tree and its manufacture into rough products. This is especially necessary with trees felled from April to September, in the region north of the Gulf States, and from March to November in the latter, while the late fall and winter cutting should all be worked up by March or April. [108]

2. If the round timber must be left in the woods or on the skidways during the danger period, every precaution should be taken to facilitate rapid drying of the inner bark, by keeping the logs off the ground in the sun, or in loose piles; or else the opposite extreme should be adopted and the logs kept in water.

3. The immediate removal of all the bark from poles, posts, and other material which will not be seriously damaged by checking or season checks.

4. To determine and utilize the proper months or seasons to girdle or fell different kinds of trees: Bald cypress in the swamps of the South are "girdled" in order that they may die, and in a few weeks or months dry out and become light enough to float. This method has been extensively adopted in sections where it is the only practicable one by which the timber can be transported to the sawmills. It is found, however, that some of these "girdled" trees are especially attractive to several species of ambrosia beetles (Figs.

22 and 23), round-headed borers (Fig. 24) and timber worms (Fig. 25), which cause serious injury to the sapwood or heartwood, while other trees "girdled" at a different time or season are not injured. This suggested to the writer the importance of experiments to determine the proper time to "girdle" trees to avoid losses, and they are now being conducted on an extensive scale by the United States Forest Service, in co-operation with prominent cypress operators in different sections of the cypress-growing region.

Saplings

Saplings, including hickory and other round hoop-poles and similiar products, are subject to serious injuries and destruction by round- and flat-headed borers (Fig. 24), and certain species of powder post borers (Figs. 26 and 27) before the bark and wood are dead or dry, and also by other powder post borers (Fig. 28) after they are dried and [109] seasoned. The conditions favoring attack by the former class are those resulting from leaving the poles in piles or bundles in or near the forest for a few weeks during the season of insect activity, and by the latter from leaving them stored in one place for several months.

Stave, Heading and Shingle Bolts

These are attacked by ambrosia beetles (Figs. 22 and 23), and the oak timber worm (Fig. 25, *a*), which, as has been frequently reported, cause serious losses. The conditions favoring attack by these insects are similiar to those mentioned under "Round Timber." The insects may enter the wood before the bolts are cut from the log or afterward, especially if the bolts are left in moist, shady places in the woods, in close piles during the danger period. If cut during the warm season, the bark should be removed and the bolts converted into the smallest practicable size and piled in such manner as to facilitate rapid drying.

Unseasoned Products in the Rough

Freshly sawn hardwood, placed in close piles during warm, damp weather in July and September, presents especially favorable conditions for injury by ambrosia beetles (Figs. 22, *a*, and 23, *a*). This is due to the continued moist condition of such material.

Heavy two-inch or three-inch stuff is also liable to attack even in loose piles with lumber or cross sticks. An example of the latter was found in a valuable lot of mahogany lumber of first grade, the value of which was reduced two thirds by injury from a native ambrosia beetle. Numerous complaints have been received from different sections of the country of this class of injury to oak, poplar, gum, and other hardwoods. In all cases it is the moist condition and retarded drying of the lumber which induces attack; therefore, any method which will provide for the rapid drying of the wood before or after piling will tend to prevent losses.

It is important that heavy lumber should, as far as possible, be cut in the winter months and piled so that it [110] will be well dried out before the middle of March. Square timber, stave and heading bolts, with the bark on, often suffer from injuries by flat- or round-headed borers, hatching from eggs deposited in the bark of the logs before they are sawed and piled. One example of serious damage and loss was reported in which white pine staves for paint buckets and other small wooden vessels, which had been sawed from small logs, and the bark left on the edges, were attacked by a round-headed borer, the adults having deposited their eggs in the bark after the stock was sawn and piled. The character of the injury is shown in Fig. 29. Another example was reported from a manufacturer in the South, where the pieces of lumber which had strips of bark on one side were seriously damaged by the same kind of borer, the eggs having been deposited in the logs before sawing or in the bark after the lumber was piled. If the eggs are deposited in the logs, and the borers have entered the inner bark or the wood before sawing, they may continue their work regardless of methods of piling, but if such lumber is cut from new logs and placed in the pile while green, with the bark surface up, it will be much less liable to attack than if piled with the bark edges down. This liability of lumber with bark edges or sides to be attacked by insects suggests the importance of the removal of the bark, to prevent damage, or, if this is not practicable, the lumber with the bark on the sides should be piled in open, loose piles with the bark up, while that with the bark on the edges should be placed on the outer edges of the piles, exposed to the light and air.

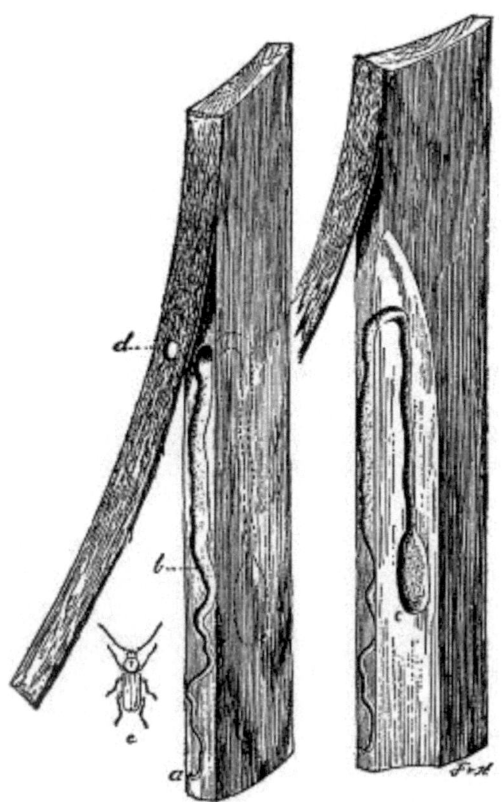

Fig. 29. Work of Round-headed Borers, *Callidium antennatum*, in White Pine Bucket Staves from New Hampshire. *a*, where egg was deposited in bark; *b*, larval mine; *c*, pupal cell; *d*, exit in bark; *e*, adult.

In the Southern States it is difficult to keep green timber in the woods or in piles for any length of time, because of the rapidity which wood-destroying fungi attack it. This is particularly true during the summer season, when the humidity is greatest. There is really no easily-applied, general specific for these summer troubles in the handling of wood, but there are some suggestions that are worth while that it may be well to mention. One of these, and the most important, is to remove all the bark from the timber that has been cut, just as soon as possible after felling. And, in this, emphasis should be laid on the ALL, [111] as a piece of bark no larger than a man's little finger will furnish an entering place for insects, and once they get in, it is a difficult matter to get rid of them, for they seldom stop boring until they ruin the stick. And again, after the timber has been felled and the bark removed, it is well to get it to the mill pond or cut up into merchantable sizes and on to the pile as soon as possible. What is wanted is to get the timber up off the ground, to a place where it can get plenty of air, to enable the sap to dry up before it sours; and, besides, large units of wood are more likely to crack open on the ends from the [112] heat than they would if cut up into the smaller units for merchandizing.

A moist condition of lumber and square timber, such as results from close or solid piles, with the bottom layers on the ground or on foundations of old decaying logs or near decaying stumps and logs, offers especially favorable conditions for the attack of white ants.

Seasoned Products in the Rough

Seasoned or dry timber in stacks or storage is liable to injury by powder post borers (Fig. 28). The conditions favoring attack are: (1) The presence of a large proportion of sapwood, as in hickory, ash, and similiar woods; (2) material which is two or more years old, or that which has been kept in one place for a long time; (3) access to old infested material. Therefore, such stock should be frequently examined for evidence of the presence of these insects. This is always indicated by fine, flour-like powder on or beneath the piles, or otherwise associated with such material. All infested material should be at once removed and the infested parts destroyed by burning.

Dry Cooperage Stock and Wooden Truss Hoops

These are especially liable to attack and serious injury by powder post borers (Fig. 28), under the same or similiar conditions as the preceding.

Staves and Heads of Barrels containing Alcoholic Liquids

These are liable to attack by ambrosia beetles (Figs. 22, *a*, and 23, *a*), which are attracted by the moist condition and possibly by the peculiar odor of the wood, resembling that of dying sapwood of trees and logs, which is their normal breeding place.

There are many examples on record of serious losses of liquors from leakage caused by the beetles boring through the staves and heads of the barrels and casks in cellars and storerooms.

The condition, in addition to the moisture of the wood, which is favorable for the presence of the beetles, is proximity [113] to their breeding places, such as the trunks and stumps of recently felled or dying oak, maple, and other hardwood or deciduous trees; lumber yards, sawmills, freshly-cut cordwood, from living or dead trees, and forests of hardwood timber. Under such conditions the beetles occur in great numbers, and if the storerooms and cellars in which the barrels are kept stored are damp, poorly ventilated, and readily accessible to them, serious injury is almost certain to follow.

WATER IN WOOD

DISTRIBUTION OF WATER IN WOOD

Local Distribution of Water in Wood

As seasoning means essentially the more or less rapid evaporation of water from wood, it will be necessary to discuss at the very outset where water is found in wood, and its local seasonal distribution in a tree.

Water may occur in wood in three conditions: (1) It forms the greater part (over 90 per cent) of the protoplasmic contents of the living cells; (2) it saturates the walls of all cells; and (3) it entirely or at least partly fills the cavities of the lifeless cells, fibres, and vessels.

In the sapwood of pine it occurs in all three forms; in the heartwood only in the second form, it merely saturates the walls.

Of 100 pounds of water associated with 100 pounds of dry wood substance taken from 200 pounds of fresh sapwood of white pine, about 35 pounds are needed to saturate the cell walls, less than 5 pounds are contained in the living cells, and the remaining 60 pounds partly fill the cavities of the wood fibres. This latter forms the sap as ordinarily understood.

The wood next to the bark contains the most water. In the species which do not form heartwood, the decrease toward the pith is gradual, but where heartwood is formed the change from a more moist to a drier condition is usually quite abrupt at the sapwood limit.

In long-leaf pine, the wood of the outer one inch of a disk may contain 50 per cent of water, that of the next, or the second inch, only 35 per cent, and that of the heartwood,

[115] only 20 per cent. In such a tree the amount of water in any one section varies with the amount of sapwood, and is greater for the upper than the lower cuts, greater for the limbs than the stems, and greatest of all in the roots.

Different trees, even of the same kind and from the same place, differ as to the amount of water they contain. A thrifty tree contains more water than a stunted one, and a young tree more than on old one, while the wood of all trees varies in its moisture relations with the season of the year.

Seasonal Distribution of Water in Wood

It is generally supposed that trees contain less water in winter than in summer. This is evidenced by the popular saying that "the sap is down in the winter." This is probably not always the case; some trees contain as much water in winter as in summer, if not more. Trees normally contain the greatest amount of water during that period when the roots are active and the leaves are not yet out. This activity commonly begins in January, February, and March, the exact time varying with the kind of timber and the local atmospheric conditions. And it has been found that green wood becomes lighter or contains less water in late spring or early summer, when transpiration through the foliage is most rapid. The amount of water at any one season, however, is doubtless much influenced by the amount of moisture in the soil. The fact that the bark peels easily in the spring depends on the presence of incomplete, soft tissue found between wood and bark during this season, and has little to do with the total amount of water contained in the wood of the stem.

Even in the living tree a flow of sap from a cut occurs only in certain kinds of trees and under special circumstances. From boards, felled timber, etc., the water does not flow out, as is sometimes believed, but must be evaporated. The seeming exceptions to this rule are mostly referable to two causes; clefts or "shakes" will [116] allow water contained in them to flow out, and water is forced out of sound wood, if very sappy, whenever the wood is warmed, just as water flows from green wood when put in a stove.

Composition of Sap

The term "sap" is an ambiguous expression. The sap in the tree descends through the bark, and except in early spring is not present in the wood of the tree except in the medullary rays and living tissues in the "sapwood."

What flows through the "sapwood" is chiefly water brought from the soil. It is not pure water, but contains many substances in solution, such as mineral salts, and in certain species—maple, birch, etc., it also contains at certain times a small percentage of sugar and other organic matter.

The water rises from the roots through the sapwood to the leaves, where it is converted into true "sap" which descends through the bark and feeds the living tissues between the bark and the wood, which tissues make the annual growth of the trunk. The wood itself contains very little true sap and the heartwood none.

The wood contains, however, mineral substances, organic acids, volatile oils and gums, as resin, cedar oil, etc.

All the conifers—pines, cedars, junipers, cypresses, sequoias, yews, and spruces—contain resin. The sap of deciduous trees—those which shed their leaves at stated seasons—is lacking in this element, and its constituents vary greatly in the different species. But there is one element common to all trees, and for that matter to almost all plant growth, and that is albumen.

Both resin and albumen, as they exist in the sap of woods, are soluble in water; and both harden with heat, much the same as the white of an egg, which is almost pure albumen.

These organic substances are the dissolved reserve food, stored during the winter in the pith rays, etc., of the wood and bark; generally but a mere trace of them is to be found.

From this it appears that the solids contained [117] in the sap, such as albumen, gum, sugar, etc., cannot exercise the influence on the strength of the wood which is so commonly claimed for them.

Effects of Moisture on Wood

The question of the effect of moisture upon the strength and stiffness of wood offers a wide scope for study, and authorities consulted differ in conclusions. Two authorities give the tensile strength in pounds per square inch for white oak as 10,000 and 19,500, respectively; for spruce, 8,000 to 19,500, and other species in similiar startling contrasts.

Wood, we are told, is composed of organic products. The chief material is cellulose, and this in its natural state in the living plant or green wood contains from 25 to 35 per cent of its weight in moisture. The moisture renders the cellulose substance pliable. What the physical action of the water is upon the molecular structure of organic material, to render it softer and more pliable, is largely a matter of conjecture.

The strength of a timber depends not only upon its relative freedom from imperfections, such as knots, crookedness of grain, decay, wormholes or ring-shakes, but also upon its density; upon the rate at which it grew, and upon the arrangement of the various elements which compose it.

The factors effecting the strength of wood are therefore of two classes: (1) Those inherent in the wood itself and which may cause differences to exist between two pieces from the same species of wood or even between the two ends of a piece, and (2) those which are foreign to the wood itself, such as moisture, oils, and heat.

Though the effect of moisture is generally temporary, it is far more important than is generally realized. So great, indeed, is the effect of moisture that under some conditions it outweighs all the other causes which effect strength, with the exception, perhaps of decided imperfections in the wood itself. [118]

The Fibre Saturation Point in Wood

Water exists in green wood in two forms: (1) As liquid water contained in the cavities of the cells or pores, and (2) as "imbibed" water intimately absorbed in the substance of which the wood is composed. The removal of the free water from the cells or pores will evidently have no effect upon the physical properties or shrinkage of the wood, but as soon as any of the "imbibed" moisture is removed from the cell walls, shrinkage begins to take place and other changes occur. The strength also begins to increase at this time.

The point where the cell walls or wood substance becomes saturated is called the "fibre saturation point," and is a very significant point in the drying of wood.

It is easy to remove the free water from woods which will stand a high temperature, as it is only necessary to heat the wood slightly above the boiling point in a closed vessel, which will allow the escape of the steam as it is formed, but will not allow dry air to come in contact with the wood, so that the surface will not become dried below its saturation point. This can be accomplished with most of the softwoods, but not as a rule with the hardwoods, as they are injured by the temperature necessary.

The chief difficulties are encountered in evaporating the "imbibed" moisture and also where the free water has to be removed through its gradual transfusion instead of boiling. As soon as the imbibed moisture begins to be extracted from any portion, shrinkage takes place and stresses are set up in the wood which tend to cause checking.

The fibre saturation point lies between moisture conditions of 25 and 30 per cent of the dry weight of the wood, depending on the species. Certain species of eucalyptus, and probably other woods, however, appear to be exceptional in this respect, in that shrinkage begins to take place

at a moisture condition of 80 to 90 per cent of the dry
weight.

SECTION VII [119]

WHAT SEASONING IS

Seasoning is ordinarily understood to mean drying. When exposed to the sun and air, the water in green wood rapidly evaporates. The rate of evaporation will depend on: (1) the kind of wood; (2) the shape and thickness of the timber; and (3) the conditions under which the wood is placed or piled.

Pieces of wood completely surrounded by air, exposed to the wind and the sun, and protected by a roof from rain and snow, will dry out very rapidly, while wood piled or packed close together so as to exclude the air, or left in the shade and exposed to rain and snow, will dry out very slowly and will also be subject to mould and decay.

But seasoning implies other changes besides the evaporation of water. Although we have as yet only a vague conception as to the exact nature of the difference between seasoned and unseasoned wood, it is very probable that one of these consists in changes in the albuminous substances in the wood fibres, and possibly also in the tannins, resins, and other incrusting substances. Whether the change in these substances is merely a drying-out, or whether it consists in a partial decomposition is at yet undetermined. That the change during the seasoning process is a profound one there can be no doubt, because experience has shown again and again that seasoned wood fibre is very much more permeable, both for liquids and gases than the living, unseasoned fibre.

One can picture the albuminous substances as forming a coating which dries out and possibly disintegrates when the wood dries. The drying-out may result in considerable shrinkage, which may make the wood fibre more porous. It is also possible that there are oxidizing influences [120] at work within these substances which result in their disintegration. Whatever the exact nature of the change may be, one can say without hesitation that exposure to the wind and air brings about changes in the wood, which are of such a nature that the wood becomes drier and more permeable.

When seasoned by exposure to live steam, similiar changes may take place; the water leaves the wood in the form of steam, while the organic compounds in the walls probably coagulate or disintegrate under the high temperature.

The most effective seasoning is without doubt that obtained by the uniform, slow drying which takes place in properly constructed piles outdoors, under exposure to the winds and the sun and under cover from the rain and snow, and is what has been termed "air-seasoning." By air-seasoning oak and similiar hardwoods, nature performs certain functions that cannot be duplicated by any artificial means. Because of this, woods of this class cannot be successfully kiln-dried green from the saw.

In drying wood, the free water within the cells passes through the cell walls until the cells are empty, while the cell walls remain saturated. When all the free water has been removed, the cell walls begin to yield up their moisture. Heat raises the absorptive power of the fibres and so aids the passage of water from the interior of the cells. A confusion in the word "sap" is to be found in many discussions of kiln-drying; in some instances it means water, in other cases it is applied to the organic substances held in a water solution in the cell cavities. The term is best confined to the organic substances from the living cell. These substances, for the most part of the nature of sugar, have a strong attraction for water and water vapor, and so retard drying and absorb moisture into dried wood. High temperatures, especially those produced by live steam, appear to destroy these organic compounds and therefore both to retard and to limit the reabsorption of moisture when the wood is subsequently exposed to the atmosphere.

Air-dried wood, under ordinary atmospheric temperatures, [121] retains from 10 to 20 per cent of moisture, whereas kiln-dried wood may have no more than 5 per cent as it comes from the kiln. The exact figures for a given species depend in the first case upon the weather conditions, and in the second case upon the temperature in the kiln and the time during which the wood is exposed to it. When wood that has been kiln-dried is allowed to stand in the open, it apparently ceases to reabsorb moisture from the air before its moisture content equals that of wood which has merely been air-dried in the same place, and under the same conditions, in other words kiln-dried wood will not absorb as much moisture as air-dried wood under the same conditions.

Difference between Seasoned and Unseasoned Wood

Although it has been known for a long time that there is a marked difference in the length of life of seasoned and of unseasoned wood, the consumers of wood have shown very little interest in its seasoning, except for the purpose of doing away with the evils which result from checking, warping, and shrinking. For this purpose both kiln-drying and air-seasoning are largely in use.

The drying of material is a subject which is extremely important to most industries, and in no industry is it of more importance than in the lumber trade. Timber drying means not only the extracting of so much water, but goes very deeply into the quality of the wood, its workability and its cell strength, etc.

Kiln-drying, which dries the wood at a uniformly rapid rate by artificially heating it in inclosed rooms, has become a part of almost every woodworking industry, as without it the construction of the finished product would often be impossible. Nevertheless much unseasoned or imperfectly seasoned wood is used, as is evidenced by the frequent shrinkage and warping of the finished articles. This is explained to a certain extent by the fact that the manufacturer is often so hard pressed for his product that he is forced to send out an inferior article, which the consumer is willing to accept in that condition rather than [122] to wait several weeks or months for an article made up of thoroughly seasoned material, and also that dry kilns are at present constructed and operated largely without thoroughgoing system.

Forms of kilns and mode of operation have commonly been copied by one woodworking plant after the example of some neighboring establishment. In this way it has been brought about that the present practices have many shortcomings. The most progressive operators, however, have experimented freely in the effort to secure special results desirable for their peculiar products. Despite the diversity of practice, it is possible to find among the larger and more enterprising operators a measure of agreement, as to both methods and results, and from this to outline the essentials of a correct theory. As a result, properly seasoned wood commands a high price, and in some cases cannot be obtained at all.

Wood seasoned out of doors, which by many is supposed to be much superior to kiln-dried material, is becoming very scarce, as the demand for any kind of wood is so great that it is thought not to pay to hold it for the time necessary to season it properly. How long this state of affairs is going to last it is difficult to say, but it is believed that a reaction will come when the consumer learns that in the long run it does not pay to use poorly seasoned material. Such a condition has now arisen in connection with another phase of the seasoning of wood; it is a commonly accepted fact that dry wood will not decay nearly so fast as wet or green wood; nevertheless, the immense superiority of seasoned over unseasoned wood for all purposes where resistance to decay is necessary has not been sufficiently recognized. In the times when wood of all kinds was both plentiful and cheap, it mattered little in most cases how long it lasted or resisted decay. Wood used for furniture, flooring, car construction, cooperage, etc., usually got some chance to dry out before or after it was placed in use. The wood which was exposed to decaying influences was generally selected from those woods which, whatever their other qualities might be, would resist decay longest. [123]

To-day conditions have changed, so that wood can no longer be used to the same extent as in former years. Inferior woods with less lasting qualities have been pressed into service. Although haphazard methods of cutting and subsequent use are still much in vogue, there are many signs that both lumbermen and consumers are awakening to the fact that such carelessness and wasteful methods of handling wood will no longer do, and must give way to more exact and economical methods. The reason why many manufacturers and consumers of wood are still using the older methods is perhaps because of long custom, and because they have not yet learned that, though the saving to be obtained by the application of good methods has at all times been appreciable, now, when wood is more valuable, a much greater saving is possible. The increased cost of applying economical methods is really very slight, and is many times exceeded by the value of the increased service which can be secured through its use.

Manner of Evaporation of Water

The evaporation of water from wood takes place largely through the ends, *i.e.*, in the direction of the longitudinal axis of the wood fibres. The evaporation from the other surfaces takes place very slowly out of doors, and with greater rapidity in a dry kiln. The rate of evaporation differs both with the kind of timber and its shape; that is, thin material will dry more rapidly than heavier stock. Sapwood dries faster than heartwood, and pine more rapidly than oak or other hardwoods.

Tests made show little difference in the rate of evaporation in sawn and hewn stock, the results, however, not being conclusive. Air-drying out of doors takes from two months to a year, the time depending on the kind of timber, its thickness, and the climatic conditions. After wood has reached an air-dry condition it absorbs water in small quantities after a rain or during damp weather, much of which is immediately lost again when a few warm, dry days follow. In this way wood exposed to the weather [124] will continue to absorb water and lose it for indefinite periods.

When soaked in water, seasoned woods absorb water rapidly. This at first enters into the wood through the cell walls; when these are soaked, the water will fill the cell lumen, so that if constantly submerged the wood may become completely filled with water.

The following figures show the gain in weight by absorption of several coniferous woods, air-dry at the start, expressed in per cent of the kiln-dry weight:

Absorption of Water by Dry Wood

	White Pine	Red Cedar	Hemlock	Tamarac
Air-dried	108	109	111	108
Kiln-dried	100	100	100	100
In water 1 day	135	120	133	129
In water 2 days	147	126	144	136
In water 3 days	154	132	149	142
In water 4 days	162	137	154	147
In water 5 days	165	140	158	150
In water 7 days	176	143	164	156
In water 9 days	179	147	168	157
In water 11 days	184	149	173	159
In water 14 days	187	150	176	159

In water 17 days	192	152	176	161
In water 25 days	198	155	180	161
In water 30 days	207	158	183	166

Rapidity of Evaporation

The rapidity with which water is evaporated, that is, the rate of drying, depends on the size and shape of the piece and on the structure of the wood. An inch board dries more than four times as fast as a four-inch plank, and more than twenty times as fast as a ten-inch timber. White pine dries faster than oak. A very moist piece of pine or oak will, during one hour, lose more than four times as much water per square inch from the cross-section, but only one half as much from the tangential as from the radial section. In a long timber, where the ends or cross-sections form but a small part of the drying surface, this difference [125] is not so evident. Nevertheless, the ends dry and shrink first, and being opposed in this shrinkage by the more moist adjoining parts, they check, the cracks largely disappearing as seasoning progresses.

High temperatures are very effective in evaporating the water from wood, no matter how humid the air, and a fresh piece of sapwood may lose weight in boiling water, and can be dried to quite an extent in hot steam.

In drying chemicals or fabrics, all that is required is to provide heat enough to vaporize the moisture and circulation enough to carry off the vapor thus secured, and the quickest and most economical means to these ends may be used. While on the other hand, in drying wood, whether in the form of standard stock or the finished product, the application of the requisite heat and circulation must be carefully regulated throughout the entire process, or warping and checking are almost certain to result. Moreover, wood of different shapes and thicknesses is very differently effected by the same treatment. Finally, the tissues composing the wood, which vary in form and physical properties, and which cross each other in regular directions, exert their own peculiar influences upon its behavior during drying. With our native woods, for instance, summer-wood and spring-wood show distinct tendencies in drying, and the same is true in a less degree of heartwood, as contrasted with sapwood. Or, again, pronounced medullary rays further complicate the drying problem.

Physical Properties that influence Drying

The principal properties which render the drying of wood peculiarly difficult are: (1) The irregular shrinkage; (2) the different ways in which water is contained; (3) the manner in which moisture transfuses through the wood from the center to the surface; (4) the plasticity of the wood substance while moist and hot; (5) the changes which take place in the hygroscopic and chemical nature of the surface; and (6) the difference produced in the total shrinkage by different rates of drying.

The shrinkage is unequal in different directions and in different portions of the same piece. It is greatest in [126] the circumferential direction of the tree, being generally twice as great in this direction as in the radial direction. In the longitudinal direction, for most woods, it is almost negligible, being from 20 to over 100 times as great circumferentially as longitudinally.

There is a great variation in different species in this respect. Consequently, it follows from necessity that large internal strains are set up when the wood shrinks, and were it not for its plasticity it would rupture. There is an enormous difference in the total amount of shrinkage of different species of wood, varying from a shrinkage of only 7 per cent in volume, based on the green dimensions, in the case of some of the cedars to nearly 50 per cent in the case of some species of eucalyptus.

When the free water in the capillary spaces of the wood fibre is evaporated it follows the laws of evaporation from capillary spaces, except that the passages are not all free passages, and much of the water has to pass out by a process of transfusion through the moist cell walls. These cell walls in the green wood completely surround the cell cavities so that there are no openings large enough to offer a passage to water or air.

The well-known "pits" in the cell walls extend through the secondary thickening only, and not through the primary walls. This statement applies to the tracheids and parenchyma cells in the conifer (gymnosperms), and to the tracheids, parenchyma cells, and the wood fibres in the broad-leaved trees (angiosperms); the vessels in the latter, however, form open passages except when clogged by ingrowth called tyloses, and the resin canals in the former sometimes form occasional openings.

By heating the wood above the boiling point, corresponding to the external pressure, the free water passes through the cell walls more readily.

To remove the moisture from the wood substance requires heat in addition to the latent heat of evaporation, because the molecules of moisture are so intimately associated with the molecules, minute particles composing the wood, that energy is required to separate them therefrom. [127]

Carefully conducted experiments show this to be from 16.6 to 19.6 calories per grain of dry wood in the case of beech, long-leaf pine, and sugar maple.

The difficulty imposed in drying, however, is not so much the additional heat required as it is in the rate at which the water transfuses through the solid wood.

SECTION VIII [128]

ADVANTAGES IN SEASONING

Three most important advantages of seasoning have already been made apparent:

1. Seasoned timber lasts much longer than unseasoned. Since the decay of timber is due to the attacks of wood-destroying fungi, and since the most important condition of the growth of these fungi is water, anything which lessens the amount of water in wood aids in its preservation.

2. In the case of treated timber, seasoning before treatment greatly increases the effectiveness of the ordinary methods of treatment, and seasoning after treatment prevents the rapid leaching out of the salts introduced to preserve the timber.

3. The saving in freight where timber is shipped from one place to another. Few persons realize how much water green wood contains, or how much it will lose in a comparatively short time. Experiments along this line with lodgepole pine, white oak, and chestnut gave results which were a surprise to the companies owning the timber.

Freight charges vary considerably in different parts of the country; but a decrease of 35 to 40 per cent in weight is important enough to deserve everywhere serious consideration from those in charge of timber operations.

When timber is shipped long distances over several roads, as is coming to be more and more the case, the saving in freight will make a material difference in the cost of lumber operations, irrespective of any other advantages of seasoning. [129]

Prevention of Checking and Splitting

Under present methods much timber is rendered unfit for use by improper seasoning. Green timber, particularly when cut during January, February, and March, when the roots are most active, contains a large amount of water. When exposed to the sun and wind or to high temperatures in a drying room, the water will evaporate more rapidly from the outer than from the inner parts of the piece, and more rapidly from the ends than from the sides. As the water evaporates, the wood shrinks, and when the shrinkage is not fairly uniform the wood cracks and splits.

When wet wood is piled in the sun, evaporation goes on with such unevenness that the timbers split and crack in some cases so badly as to become useless for the purpose for which it was intended. Such uneven drying can be prevented by careful piling, keeping the logs immersed in a log pond until wanted, or by piling or storing under an open shed so that the sun cannot get at them.

Experiments have also demonstrated that injury to stock in the way of checking and splitting always develops immediately after the stock is taken into the dry kiln, and is due to the degree of humidity being too low.

The receiving end of the kiln should always be kept moist, where the stock has not been steamed before being put into the kiln, as when the air is too dry it tends to dry the outside of the stock first—which is termed "case-hardening"—and in so doing shrinks and closes up the pores. As the material is moved down the kiln (as in the case of "progressive kilns"), it absorbs a continually increasing amount of heat, which tends to drive off the moisture still present in the center of the piece, the pores on the outside having been closed up, there is no exit for the vapor or steam that is being rapidly formed in the center of the piece. It must find its way out in some manner, and in doing so sets up strains, which result either in checking or splitting. If the humidity had been kept higher, the outside of the piece would not have dried so quickly, and the pores would have remained [130] open for the exit of the moisture from the interior of the piece, and this trouble would have been avoided. (See also article following.)

Shrinkage of Wood

Since in all our woods, cells with thick walls and cells with thin walls are more or less intermixed, and especially as the spring-wood and summer-wood nearly always differ from each other in this respect, strains and tendencies to warp are always active when wood dries out, because the summer-wood shrinks more than the spring-wood, and heavier wood in general shrinks more than light wood of the same kind.

If a thin piece of wood after drying is placed upon a moist surface, the cells on the under side of the piece take up moisture and swell before the upper cells receive any moisture. This causes the under side of the piece to become longer than the upper side, and as a consequence warping occurs. Soon, however, the moisture penetrates to all the cells and the piece straightens out. But while a thin board of pine curves laterally it remains quite straight lengthwise, since in this direction both shrinkage and swelling are small. If one side of a green board is exposed to the sun, warping is produced by the removal of water and consequent shrinkage of the side exposed; this may be eliminated by the frequent turning of the topmost pieces of the piles in order that they may be dried evenly.

As already stated, wood loses water faster from the ends than from the longitudinal faces. Hence the ends shrink at a different rate from the interior parts. The faster the drying at the surface, the greater is the difference in the moisture of the different parts, and hence the greater the strains and consequently also the greater amount of checking. This becomes very evident when freshly cut wood is placed in the sun, and still more when put into a hot, dry kiln. While most of these smaller checks are only temporary, closing up again, some large radial checks remain and even grow larger as drying progresses. Their cause is a different one and will presently be explained. The temporary checks not only appear at the ends, but [131] are developed on the sides also, only to a much smaller degree. They become especially annoying on the surface of thick planks of hard-woods, and also on peeled logs when exposed to the sun.

So far we have considered the wood as if made up only of parallel fibres all placed longitudinally in the log. This, however, is not the case. A large part of the wood is formed by the medullary or pith rays. In pine over 15,000 of these occur on a square inch of a tangential section, and even in oak the very large rays, which are readily visible to the eye, represent scarcely a hundredth part of the number which a microscope reveals, as the cells of these rays have their length at right angles to the direction of the wood fibres.

If a large pith ray of white oak is whittled out and al-lowed to dry, it is found to shrink greatly in its width, while, as we have stated, the fibres to which the ray is firm-ly grown in the wood do not shrink in the same direction. Therefore, in the wood, as the cells of the pith ray dry they pull on the longitudinal fibres and try to shorten them, and, being opposed by the rigidity of the fibres, the pith ray is greatly strained. But this is not the only strain it has to bear. Since the fibres shrink as much again as the pith ray, in this its longitudinal direction, the fibres tend to shorten the ray, and the latter in opposing this prevents the former from shrinking as much as they otherwise would.

Thus the structure is subjected to two severe strains at right angles to each other, and herein lies the greatest diffi-culty of wood seasoning, for whenever the wood dries rap-idly these fibres have not the chance to "give" or accommo-date themselves, and hence fibres and pith rays separate and checking results, which, whether visible or not, are detrimental in the use of the wood.

The contraction of the pith rays parallel to the length of the board is probably one of the causes of the small amount of longitudinal shrinkage which has been observed in boards. This smaller shrinkage of the pith rays along the radius of the log (the length of the pith ray), opposing the shrinkage of the fibres in this direction, becomes [132] one of the causes of the second great trouble in wood seasoning, namely, the difference in the shrinkage along the radius and that along the rings or tangent. This greater tangential shrinkage appears to be due in part to the causes just mentioned, but also to the fact that the greatly shrinking bands of summer-wood are interrupted along the radius by as many bands of porous spring-wood, while they are continuous in the tangential direction. In this direction, therefore, each such band tends to shrink, as if the entire piece were composed of summer-wood, and since the summer-wood represents the greater part of the wood substance, this greater tendency to tangential shrinkage prevails.

The effect of this greater tangential shrinkage effects every phase of woodworking. It leads to permanent checks and causes the log or piece to split open on drying. Sawed in two, the flat sides of the log become convex; sawed into timber, it checks along the median line of the four faces, and if converted into boards, the latter checks considerably from the end through the center, all owing to the greater tangential shrinkage of the wood.

Briefly, then, shrinkage of wood is due to the fact that the cell walls grow thinner on drying. The thicker cell walls and therefore the heavier wood shrinks most, while the water in the cell cavities does not influence the volume of the wood.

Owing to the great difference of cells in shape, size, and thickness of walls, and still more in their arrangement, shrinkage is not uniform in any kind of wood. This irregularity produces strains, which grow with the difference between adjoining cells and are greatest at the pith rays. These strains cause warping and checking, but exist even where no outward signs are visible. They are greater if the wood is dried rapidly than if dried slowly, but can never be entirely avoided.

Temporary checks are caused by the more rapid drying of the outer parts of any stick; permanent checks are due to the greater shrinkage, tangentially, along the rings than along the radius. This, too, is the cause of most of the ordinary phenomena of shrinkage, such as [133] the difference in behavior of the entire and quartered logs, "bastard" (tangent) and rift (radial) boards, etc., and explains many of the phenomena erroneously attributed to the influence of bark, or of the greater shrinkage of outer and inner parts of any log.

Once dry, wood may be swelled again to its original size by soaking in water, boiling, or steaming. Soaked pieces on drying shrink again as before; boiled and steamed pieces do the same, but to a slightly less degree. Neither hygroscopicity, *i.e.*, the capacity of taking up water, nor shrinkage of wood can be overcome by drying at temperatures below 200 degrees Fahrenheit. Higher temperatures, however, reduce these qualities, but nothing short of a coaling heat robs wood of the capacity to shrink and swell.

Rapidly dried in a kiln, the wood of oak and other hardwoods "case-harden," that is, the outer part dries and shrinks before the interior has a chance to do the same, and thus forms a firm shell or case of shrunken, commonly checked wood around the interior. This shell does not prevent the interior from drying, but when this drying occurs the interior is commonly checked along the medullary rays, commonly called "honeycombing" or "hollow-horning." In practice this occurrence can be prevented by steaming or sweating the wood in the kiln, and still better by drying the wood in the open air or in a shed before placing in the kiln. Since only the first shrinkage is apt to check the wood, any kind of lumber which has once been air-dried (three to six months for one-inch stuff) may be subjected to kiln heat without any danger from this source.

Kept in a bent or warped condition during the first shrinkage, the wood retains the shape to which it has been bent and firmly opposes any attempt at subsequent straightening.

Sapwood, as a rule, shrinks more than heartwood of the same weight, but very heavy heartwood may shrink more than lighter sapwood. The amount of water in wood is no criterion of its shrinkage, since in wet wood most of the water is held in the cavities, where it has no effect on the volume. [134]

The wood of pine, spruce, cypress, etc., with its very regular structure, dries and shrinks evenly, and suffers much less in seasoning than the wood of broad-leaved (hardwood) trees. Among the latter, oak is the most difficult to dry without injury.

Desiccating the air with certain chemicals will cause the wood to dry, but wood thus dried at 80 degrees Fahrenheit will still lose water in the kiln. Wood dried at 120 degrees Fahrenheit loses water still if dried at 200 degrees Fahrenheit, and this again will lose more water if the temperature be raised, so that *absolutely dry wood* cannot be obtained, and chemical destruction sets in before all the water is driven off.

On removal from the kiln, the dry wood at once takes up moisture from the air, even in the driest weather. At first the absorption is quite rapid; at the end of a week a short piece of pine, 11/2 inches thick, has regained two thirds of, and, in a few months, all the moisture which it had when air-dry, 8 to 10 per cent, and also its former dimensions. In thin boards all parts soon attain the same degree of dryness. In heavy timbers the interior remains more moist for many months, and even years, than the exterior parts. Finally an equilibrium is reached, and then only the outer parts change with the weather.

With kiln-dried woods all parts are equally dry, and when exposed, the moisture coming from the air must pass through the outer parts, and thus the order is reversed. Ordinary timber requires months before it is at its best. Kiln-dried timber, if properly handled, is prime at once.

Dry wood if soaked in water soon regains its original volume, and in the heartwood portion it may even surpass it; that is to say, swell to a larger dimension than it had when green. With the soaking it continues to increase in weight, the cell cavities filling with water, and if left many months all pieces sink. Yet after a year's immersion a piece of oak 2 by 2 inches and only 6 inches long still contains air; *i.e.*, it has not taken up all the water it can. By rafting or prolonged immersion, wood loses some of its weight, soluble materials being leached [135] out, but it is not impaired either as fuel or as building material. Immersion, and still more boiling and steaming, reduce the hygroscopicity of wood and therefore also the troublesome "working," or shrinking and swelling.

Exposure in dry air to a temperature of 300 degrees Fahrenheit for a short time reduces but does not destroy the hygroscopicity, and with it the tendency to shrink and swell. A piece of red oak which has been subjected to a temperature of over 300 degrees Fahrenheit still swells in hot water and shrinks in a dry kiln.

Expansion of Wood

It must not be forgotten that timber, in common with every other material, expands as well as contracts. If we extract the moisture from a piece of wood and so cause it to shrink, it may be swelled to its original volume by soaking it in water, but owing to the protection given to most timber in dwelling-houses it is not much affected by wet or damp weather. The shrinkage is more apparent, more lasting, and of more consequence to the architect, builder, or owner than the slight expansion which takes place, as, although the amount of moisture contained in wood varies with the climate conditions, the consequence of dampness or moisture on good timber used in houses only makes itself apparent by the occasional jamming of a door or window in wet or damp weather.

Considerable expansion, however, takes place in the wood-paving of streets, and when this form of paving was in its infancy much trouble occurred owing to all allowances not having been made for this contingency, the trouble being doubtless increased owing to the blocks not being properly seasoned; curbing was lifted or pushed out of line and gully grids were broken by this action. As a rule in street paving a space of one or two inches wide is now left next to the curb, which is filled with sand or some soft material, so that the blocks may expand longitudinally without injuring the contour or affecting the curbs. But even with this arrangement it is not at all unusual for an inch or more to have to be cut off paving blocks parallel to the channels some time after the paving has [136] been laid, owing to the expansion of the wood exceeding the amounts allowed.

Considerable variation occurs in the expansion of wood blocks, and it is noticeable in the hardwoods as well as in the softwoods, and is often greater in the former than in the latter.

Expansion takes place in the direction of the length of the blocks as they are laid across the street, and causes no trouble in the other direction, the reason being that the lengthway of a block of wood is across the grain, of the timber, and it expands or contracts as a plank does. On one occasion, in a roadway forty feet wide, expansion occurred until it amounted to four inches on each side, or eight inches in all. This continual expansion and contraction is doubtless the cause of a considerable amount of wood street-paving bulging and becoming filled with ridges and depressions.

Elimination of Stain and Mildew

A great many manufacturers, and particularly those located in the Southern States, experience a great amount of difficulty in their timber becoming stained and mildewed. This is particularly true with gum wood, as it will frequently stain and mould in twenty-four hours, and they have experienced so much of this trouble that they have, in a great many instances, discontinued cutting it during the summer season.

If this matter were given proper attention they should be able to eliminate a great deal of this difficulty, as no doubt they will find after investigation that the mould has been caused by the stock being improperly piled to the weather.

Freshly sawn wood, placed in close piles during warm, damp weather in the months of July and August, presents especially favorable conditions for mould and stain. In all cases it is the moist condition and retarded drying of the wood which causes this. Therefore, any method which will provide for the rapid drying of the wood before or after piling will tend to prevent the difficulty, and the best method for eliminating mould is (1) to provide for [137] as little delay as possible between the felling of the tree, and its manufacture into rough products before the sap has had an opportunity of becoming sour. This is especially necessary with trees felled from April to September, in the region north of the Gulf States, and from March to November in the latter, while the late fall and winter cutting should all be worked up by March or April. (2) The material should be piled to the weather immediately after being sawn or cut, and every precaution should be taken in piling to facilitate rapid drying, by keeping the piles or ricks up off the ground. (3) All weeds (and emphasis should be placed on the ALL) and other vegetation should be kept well clear of the piles, in order that the air may have a clear and unobstructed passage through and around the piles, and (4) the piles should be so constructed that each stick or piece will have as much air space about it as it is possible to give to it.

If the above instructions are properly carried out, there will be little or no difficulty experienced with mould appearing on the lumber.

SECTION IX [138]

DIFFICULTIES OF DRYING WOOD

Seasoning and kiln-drying is so important a process in the manufacture of woods that a need is keenly felt for fuller information regarding it, based upon scientific study of the behavior of various species at different mechanical temperatures and under different mechanical drying processes. The special precautions necessary to prevent loss of strength or distortion of shape render the drying of wood especially difficult.

All wood when undergoing a seasoning process, either natural (by air) or mechanical (by steam or heat in a dry kiln), checks or splits more or less. This is due to the uneven drying-out of the wood and the consequent strains exerted in opposite directions by the wood fibres in shrinking. This shrinkage, it has been proven, takes place both end-wise and across the grain of the wood. The old tradition that wood does not shrink end-wise has long since been shattered, and it has long been demonstrated that there is an end-wise shrinkage.

In some woods it is very light, while in others it is easily perceptible. It is claimed that the average end shrinkage, taking all the woods, is only about 11/2 per cent. This, however, probably has relation to the average shrinkage on ordinary lumber as it is used and cut and dried. Now if we depart from this and take veneer, or basket stock, or even stave bolts where they are boiled, causing swelling both end-wise and across the grain or in dimension, after they are thoroughly dried, there is considerably more evidence of end shrinkage. In other words, a slack barrel stave of elm, say, 28 or 30 inches in length, after being [139] boiled might shrink as much in thoroughly drying-out as compared to its length when freshly cut, as a 12-foot elm board.

It is in cutting veneer that this end shrinkage becomes most readily apparent. In trimming with scoring knives it is done to exact measure, and where stock is cut to fit some specific place there has been observed a shrinkage on some of the softer woods, like cottonwood, amounting to fully 1/8 of an inch in 36 inches. And at times where drying has been thorough the writer has noted a shrinkage of 1/8 of an inch on an ordinary elm cabbage-crate strip 36 inches long, sawed from the log without boiling.

There are really no fixed rules of measurement or allowance, however, because the same piece of wood may vary under different conditions, and, again, the grain may cross a little or wind around the tree, and this of itself has a decided effect on the amount of what is termed "end shrinkage."

There is more checking in the wood of the broad-leaf (hardwood) trees than in that of the coniferous (softwood) trees, more in sapwood than in heartwood, and more in summer-wood than in spring-wood.

Inasmuch as under normal conditions of weather, water evaporates less rapidly during the early seasoning of winter, wood that is cut in the autumn and early winter is considered less subject to checking than that which is cut in spring and summer.

Rapid seasoning, except after wood has been thoroughly soaked or steamed, almost invariably results in more or less serious checking. All hardwoods which check or warp badly during the seasoning should be reduced to the smallest practicable size before drying to avoid the injuries involved in this process, and wood once seasoned *should never again be exposed to the weather*, since all injuries due to seasoning are thereby aggravated.

Seasoning increases the strength of wood in every respect, and it is therefore of great importance to protect the wood against moisture. [140]

Changes rendering Drying difficult

An important property rendering drying of wood peculiarly difficult is the changes which occur in the hygroscopic properties of the surface of a stick, and the rate at which it will allow moisture to pass through it. If wood is dried rapidly the surface soon reaches a condition where the transfusion is greatly hindered and sometimes appears almost to cease. The nature of this action is not well understood and it differs greatly in different species. Bald cypress (*Taxodium distichum*) is an example in which this property is particularly troublesome. The difficulty can be overcome by regulating the humidity during the drying operation. It is one of the factors entering into production of what is called "case-hardening" of wood, where the surface of the piece becomes hardened in a stretched or expanded condition, and subsequent shrinkage of the interior causes "honeycombing," "hollow-horning," or internal checking. The outer surface of the wood appears to undergo a chemical change in the nature of hydrolization or oxidization, which alters the rate of absorption and evaporation in the air.

As the total amount of shrinkage varies with the rate at which the wood is dried, it follows that the outer surface of a rapidly dried board shrinks less than the interior. This sets up an internal stress, which, if the board be afterward resawed into two thinner boards by slicing it through the middle, causes the two halves to cup with their convex surfaces outward. This effect may occur even though the moisture distribution in the board has reached a uniform condition, and the board is thoroughly dry before it is resawed. It is distinct from the well-known "case-hardening" effect spoken of above, which is caused by unequal moisture conditions.

The manner in which the water passes from the interior of a piece of wood to its surface has not as yet been fully determined, although it is one of the most important factors which influence drying. This must involve a transfusion of moisture through the cell walls, since, as already mentioned, except for the open vessels in the hardwoods, [141] free resin ducts in the softwoods, and possibly the intercellular spaces, the cells of green wood are enclosed by membranes and the water must pass through the walls or the membranes of the pits. Heat appears to increase this transfusion, but experimental data are lacking.

It is evident that to dry wood properly a great many factors must be taken into consideration aside from the mere evaporation of moisture.

Losses Due to Improper Kiln-drying

In some cases there is practically no loss in drying, but more often it ranges from 1 to 3 per cent, and 7 to 10 per cent in refractory woods such as gum. In exceptional instances the losses are as high as 33 per cent.

In air-drying there is little or no control over the process; it may take place too rapidly on some days and too slowly on others, and it may be very non-uniform.

Hardwoods in large sizes almost invariably check.

By proper kiln-drying these unfavorable circumstances may be eliminated. However, air-drying is unquestionably to be preferred to bad kiln-drying, and when there is any doubt in the case it is generally safer to trust to air-drying.

If the fundamental principles are all taken care of, green lumber can be better dried in the dry kiln.

Properties of Wood that affect Drying

It is clear, from the previous discussion of the structure of wood, that this property is of first importance among those influencing the seasoning of wood. The free water way usually be extracted quite readily from porous hardwoods. The presence of tyloses in white oak makes even this a difficult problem. On the other hand, its more complex structure usually renders the hygroscopic moisture quite difficult to extract.

The lack of an open, porous structure renders the transfusion of moisture through some woods very slow, while the reverse may be true of other species. The point of interest is that all the different variations in structure [142] affect the drying rates of woods. The structure of the gums suggests relatively easy seasoning.

Shrinkage is a very important factor affecting the drying of woods. Generally speaking, the greater the shrinkage the more difficult it is to dry wood. Wood shrinks about twice as much tangentially as radially, thus introducing very serious stresses which may cause loss in woods whose total shrinkage is large. It has been found that the amount of shrinkage depends, to some extent, on the rate and temperature at which woods season. Rapid drying at high or low temperature results in slight shrinkage, while slow drying, especially at high temperature, increases the shrinkage.

As some woods must be dried in one way and others in other ways, to obtain the best general results, this effect may be for the best in one case and the reverse in others. As an example one might cite the case of Southern white oak. This species must be dried very slowly at low temperatures in order to avoid the many evils to which it is heir. It is interesting to note that this method tends to increase the shrinkage, so that one might logically expect such treatment merely to aggravate the evils. Such is not the case, however, as too fast drying results in other defects much worse than that of excessive shrinkage.

Thus we see that the shrinkage of any given species of wood depends to a great extent on the method of drying. Just how much the shrinkage of gum is affected by the temperature and drying rate is not known at present. There is no doubt that the method of seasoning affects the shrinkage of the gums, however. It is just possible that these woods may shrink longitudinally more than is normal, thus furnishing another cause for their peculiar action under certain circumstances. It has been found that the properties of wood which affect the seasoning of the gums are, in the order of their importance: (1) The indeterminate and erratic grain; (2) the uneven shrinkage with the resultant opposing stresses; (3) the plasticity under high temperature while moist; and (4) the slight apparent lack of cohesion between the fibres. The first, second, and fourth properties are clearly detrimental, [143] while the third may possibly be an advantage in reducing checking and "case-hardening."

The grain of the wood is a prominent factor also affecting the problem. It is this factor, coupled with uneven shrinkage, which is probably responsible, to a large extent, for the action of the gums in drying. The grain may be said to be more or less indeterminate. It is usually spiral, and the spiral may reverse from year to year of the tree's growth. When a board in which this condition exists begins to shrink, the result is the development of opposing stresses, the effect of which is sometimes disastrous. The shrinkage around the knots seems to be particularly uneven, so that checking at the knots is quite common.

Some woods, such as Western red cedar, redwood, and eucalyptus, become very plastic when hot and moist. The result of drying-out the free water at high temperature may be to collapse the cells. The gums are known to be quite soft and plastic, if they are moist, at high temperature, but they do not collapse so far as we have been able to determine.

The cells of certain species of wood appear to lack cohesion, especially at the junction between the annual rings. As a result, checks and ring shakes are very common in Western larch and hemlock. The parenchyma cells of the medullary rays in oak do not cohere strongly and often check open, especially when steamed too severely.

Unsolved Problems in Kiln-drying

1. Physical data of the properties of wood in relation to heat are meagre.

2. Figures on the specific heat of wood are not readily available, though upon this rests not only the exact opera-

tion of heating coils for kilns, but the theory of kiln-drying as a whole.

3. Great divergence is shown in the results of experiments in the conductivity of wood. It remains to be seen whether the known variation of conductivity with moisture content will reduce these results to uniformity. [144]

4. The maximum or highest temperature to which the different species of wood may be exposed without serious loss of strength has not yet been determined.

5. The optimum or absolute correct temperature for drying the different species of wood is as yet entirely unsettled.

6. The inter-relation between wood and water is as imperfectly known to dry-kiln operators as that between wood and heat.

7. What moisture conditions obtain in a stick of air-dried wood?

8. How is the moisture distinguished?

9. What is its form?

10. What is the meaning of the peculiar surface conditions which even in air-dried wood appear to indicate incipient "case-hardening"?

11. The manner in which the water passes from the interior of a piece of wood to its surface has not as yet been fully determined.

These questions can be answered thus far only by speculation or, at best, on the basis of incomplete data.

Until these problems are solved, kiln-drying must necessarily remain without the guidance of complete scientific theory.

A correct understanding of the principles of drying is rare, and opinions in regard to the subject are very diverse. The same lack of knowledge exists in regard to dry kilns. The physical properties of the wood which complicate the drying operation and render it distinct from that of merely evaporating free water from some substance like a piece of cloth must be studied experimentally. It cannot well be worked out theoretically.

SECTION X [145]

HOW WOOD IS SEASONED

Methods of Drying

The choice of a method of drying depends largely upon the object in view. The principal objects may be grouped under three main heads, as follows:

- 1. To reduce shipping weight.
- 2. To reduce the quantity necessary to carry in stock.
- 3. To prepare the wood for its ultimate use and improve its qualities.

When wood will stand the temperature without excessive checking or undue shrinkage or loss in strength, the first object is most readily attained by heating the wood above the boiling point in a closed chamber, with a large circulation of air or vapor, so arranged that the excess steam produced will escape. This process manifestly does not apply to many of the hardwoods, but is applicable to many of the softwoods. It is used especially in the northwestern part of the United States, where Douglas fir boards one inch thick are dried in from 40 to 65 hours, and sometimes in as short a time as 24 hours. In the latter case superheated steam at 300 degrees Fahrenheit was forced into the chamber but, of course, the lumber could not be heated thereby much above the boiling point so long as it contained any free water.

This lumber, however, contained but 34 per cent moisture to start with, and the most rapid rate was 1.6 per cent loss per hour.

The heat of evaporation may be supplied either by superheated steam or by steam pipes within the kiln itself.

The quantity of wood it is necessary to carry in stock [146] is naturally reduced when either of the other two objects is attained and, therefore, need not necessarily be discussed.

In drying to prepare for use and to improve quality, careful and scientific drying is called for. This applies more particularly to the hardwoods, although it may be required for softwoods also.

Drying at Atmospheric Pressure

Present practice of kiln-drying varies tremendously and there is no uniformity or standard method.

Temperatures vary anywhere from 65 to 165 degrees Fahrenheit, or even higher, and inch boards three to six months on the sticks are being dried in from four days to three weeks, and three-inch material in from two to five months.

All methods in use at atmospheric pressure may be classified under the following headings. The kilns may be either progressive or compartment, and preliminary steaming may or may not be used with any one of these methods:

- 1. Dry air heated. This is generally obsolete.
- 2. Moist air.

  -
    - *a.* Ventilated.
    - *b.* Forced draft.
    - *c.* Condensing.
    - *d.* Humidity regulated.
    - *e.* Boiling.
- 3. Superheated steam.

Drying under Pressure and Vacuum

Various methods of drying wood under pressures other than atmospheric have been tried. Only a brief mention of this subject will be made. Where the apparatus is available probably the quickest way to dry wood is first to heat it in saturated steam at as high a temperature as the species can endure without serious chemical change until the heat has penetrated to the center, then follow this with a vacuum. [147]

By this means the self-contained specific heat of the wood and the water is made available for the evaporation, and the drying takes place from the inside outwardly, just the reverse of that which occurs by drying by means of external heat.

When the specimen has cooled this process is then to be repeated until it has dried down to fibre-saturation point. It cannot be dried much below this point by this method, since the absorption during the heating operation will then equal the evaporation during the cooling. It may be carried further, however, by heating in partially humidified air, proportioning the relative humidity each time it is heated to the degree of moisture present in the wood.

The point to be considered in this operation is that during the heating process no evaporation shall be allowed to take place, but only during the cooling. In this way surface drying and "case-hardening" are prevented since the heat is from within and the moisture passes from the inside outwardly. However, with some species, notably oak, surface cracks appear as a network of fine checks along the medullary rays.

In the first place, it should be borne in mind that it is the heat which produces evaporation and not the air nor any mysterious property assigned to a "vacuum."

For every pound of water evaporated at ordinary temperatures approximately 1,000 British thermal units of heat are used up, or "become latent," as it is called. This is true whether the evaporation takes place in a vacuum or under a moderate air pressure. If this heat is not supplied from an outside source it must be supplied by the water itself (or the material being dried), the temperature of which will consequently fall until the surrounding space becomes saturated with vapor at a pressure corresponding to the temperature which the water has reached; evaporation will then cease. The pressure of the vapor in a space saturated with water vapor increases rapidly with increase of temperature. At a so-called vacuum of 28 inches, which is about the limit in commercial operations, and in reality signifies an actual pressure of 2 inches [148] of mercury column, the space will be saturated with vapor at 101 degrees Fahrenheit. Consequently, no evaporation will take place in such a vacuum unless the water be warmer than 101 degrees Fahrenheit, provided there is no air leakage. The qualification in regard to air is necessary, for the sake of exactness, for the following reason: In any given space the total actual pressure is made up of the combined pressures of all the gases present. If the total pressure ("vacuum") is 2 inches, and there is no air present, it is all produced by the water vapor (which saturates the space at 101 degrees Fahrenheit); but if some air is present and the total pressure is still maintained at 2 inches, then there must be less vapor present, since the air is producing part of the pressure and the space is no longer saturated at the given temperature. Consequently further evaporation may occur, with a corresponding lowering of the temperature of the water, until a balance is again reached. Without further explanation it is easy to see that but little water can be evaporated by a vacuum alone without addition of heat, and that the prevalent idea that a vacuum can of itself produce evaporation is a fallacy. If heat be supplied to the water, however, either by conduction or radiation, evaporation will take place in direct proportion to the amount of heat supplied, so long as the pressure is kept down by the vacuum pump.

At 30 inches of mercury pressure (one atmosphere) the space becomes saturated with vapor and equilibrium is established at 212 degrees Fahrenheit. If heat be now supplied to the water, however, evaporation will take place in proportion to the amount of heat supplied, so long as the pressure remains that of one atmosphere, just as in the case of the vacuum. Evaporation in this condition, where the vapor pressure at the temperature of the water is equal to the gas pressure on the water, is commonly called "boiling," and the saturated vapor entirely displaces the air under continuous operation. Whenever the space is not saturated with vapor, whether air is present or not, evaporation will take place, by boiling if no air be present or by diffusion under the presence [149] of air, until an equilibrium between temperature and vapor pressure is resumed.

Relative humidity is simply the ratio of the actual vapor pressure present in a given space to the vapor pressure when the space is saturated with vapor at the given temperature. It matters not whether air be present or not. One hundred per cent humidity means that the space contains all the vapor which it can hold at the given temperature — it is saturated. Thus at 100 per cent humidity and 212 degrees Fahrenheit the space is saturated, and since the pressure of saturated vapor at this temperature is one atmosphere, no air can be present under these conditions. If, however, the total pressure at this temperature were 20 pounds (5 pounds gauge), then it would mean that there was 5 pounds air pressure present in addition to the vapor, yet the space would still be saturated at the given temperature. Again, if the temperature were 101 degrees Fahrenheit, the pressure of saturated vapor would be only 1 pound, and the additional pressure of 14 pounds, if the total pressure were atmospheric, would be made up of air. In order to have no air present and the space still saturated at 101 degrees Fahrenheit, the total pressure must be reduced to 1 pound by a vacuum pump. Fifty per cent relative humidity, therefore, signifies that only half the amount of vapor required to saturate the space at the given temperature is present. Thus at 212 degrees Fahrenheit temperature the vapor pressure would only be 71⁄2pounds (vacuum of 15 inches gauge). If the total pressure were atmospheric, then the additional 71⁄2 pounds would be simply air.

"Live steam" is simply water-saturated vapor at a pressure usually above atmospheric. We may just as truly have live steam at pressures less than atmospheric, at a vacuum of 28 inches for instance. Only in the latter case its temperature would be lower, *viz.*, 101 degrees Fahrenheit.

Superheated steam is nothing more than water vapor at a relative humidity less than saturation, but is usually considered at pressures above atmospheric, and in the absence of air. The atmosphere at, say, 50 per cent relative [150] humidity really contains superheated steam or vapor, the only difference being that it is at a lower temperature and pressure than we are accustomed to think of in speaking of superheated steam, and it has air mixed with it to make up the deficiency in pressure below the atmosphere.

Two things should now be clear; that evaporation is produced by heat and that the presence or absence of air does not influence the amount of evaporation. It does, however, influence the rate of evaporation, which is retarded by the presence of air. The main things influencing evaporation are, first, the quantity of heat supplied and, second, the relative humidity of the immediately surrounding space.

Drying by Superheated Steam

What this term really signifies is simply water vapor in the absence of air in a condition of less than saturation. Kilns of this type are, properly speaking, vapor kilns, and usually operate at atmospheric pressure, but may be used at greater pressures or at less pressures. As stated before, the vapor present in the air at any humidity less than saturation is really "superheated steam," only at a lower pressure than is ordinarily understood by this term, and mixed with air. The main argument in favor of this process seems to be based on the idea that steam is moist heat. This is true, however, only when the steam is near saturation. When it is superheated it is just as dry as air containing the same relative humidity. For instance, steam at atmospheric pressure and heated to 248 degrees Fahrenheit has a relative humidity of only 50 per cent and is just as dry as air containing the same humidity. If heated to 306 degrees Fahrenheit, its relative humidity is reduced to 20 per cent; that is to say, the ratio of its actual vapor pressure (one atmosphere) to the pressure of saturated vapor at this temperature (five atmospheres) is 1:5, or 20 per cent. Superheated vapor in the absence of air, however, parts with its heat with great rapidity and finally becomes saturated when it has lost all of its ability to cause evaporation. In this respect it is more moist than air when it comes in contact with bodies which [151] are at a lower temperature. When saturated steam is used to heat the lumber it can raise the temperature of the latter to its own temperature, but cannot produce evaporation unless, indeed, the pressure is varied. Only by the heat supplied above the temperature of saturation can evaporation be produced.

Impregnation Methods

Methods of partially overcoming the shrinkage by impregnation of the cell walls with organic materials closely allied to the wood substance itself are in use. In one of these which has been patented, sugar is used as the impregnating material, which is subsequently hardened or "caramelized" by heating. Experiments which the United States Forest Service has made substantiate the claims that the sugar does greatly reduce the shrinkage of the wood; but the use of impregnation processes is determined rather from a financial economic standpoint than by the physical result obtained.

Another process consists in passing a current of electricity through the wet boards or through the green logs before sawing. It is said that the ligno cellulose and the sap are thus transformed by electrolysis, and that the wood subsequently dries more rapidly.

Preliminary Treatments

In many dry kiln operations, especially where the kilns are not designed for treatments with very moist air, the wood is allowed to air-season from several months to a year or more before running it into the dry kiln. In this way the surface dries below its fibre-saturation point and becomes hardened or "set" and the subsequent shrinkage is not so great. Moreover, there is less danger of surface checking in the kiln, since the surface has already passed the danger point. Many woods, however, check severely in air-drying or case-harden in the air. It is thought that such woods can be satisfactorily handled in a humidity-regulated kiln direct from the saw.

Preliminary steaming is frequently used to moisten the surface if case-hardened, and to heat the lumber through [152] to the center before drying begins. This is sometimes done in a separate chamber, but more often in a compartment of the kiln itself, partitioned off by means of a curtain which can be raised or lowered as circumstances require. This steaming is usually conducted at atmospheric pressure and frequently condensed steam is used at temperatures far below 212 degrees Fahrenheit. In a humidity-regulated kiln this preliminary treatment may be omitted, since nearly saturated conditions can be maintained and graduated as the drying progresses.

Recently the process of steaming at pressures up to 20 pounds gauge in a cylinder for short periods of time, varying from 5 to 20 minutes, is being advocated in the United States. The truck load is run into the cylinder, steamed, and then taken directly out into the air. It may subsequently be placed in the dry kiln if further drying is desired. The self-contained heat of the wood evaporates considerable moisture, and the sudden drying of the boards causes the shrinkage to be reduced slightly in some cases. Such short periods of steaming under 20 pounds pressure do not appear to injure the wood mechanically, although they do darken the color appreciably, especially of the sapwood of the species having a light-colored sap, as black walnut (*Ju-glans nigra*) and red gum (*Liquidamber styraciflua*). Longer periods of steaming have been found to weaken the wood. There is a great difference in the effect on different species, however.

Soaking wood for a long time before drying has been practised, but experiments indicate that no particularly beneficial results, from the drying standpoint, are attained thereby. In fact, in some species containing sugars and allied substances it is probably detrimental from the shrinkage standpoint. If soaked in boiling water some species shrink and warp more than if dried without this treatment.

In general, it may be said that, except possibly for short-period steaming as described above, steaming and soaking hardwoods at temperatures of 212 degrees Fahrenheit or over should be avoided if possible. [153]

It is the old saying that wood put into water shortly after it is felled, and left in water for a year or more, will be perfectly seasoned after a short subsequent exposure to the air. For this reason rivermen maintain that timber is made better by rafting. Herzenstein says: "Floating the timber down rivers helps to wash out the sap, and hence must be considered as favorable to its preservation, the more so as it enables it to absorb more preservative."

Wood which has been buried in swamps is eagerly sought after by carpenters and joiners, because it has lost all tendency to warp and twist. When first taken from the swamp the long-immersed logs are very much heavier than water, but they dry with great rapidity. A cypress log from the Mississippi Delta, which two men could barely handle at the time it was taken out some years ago, has dried out so much since then that to-day one man can lift it with ease. White cedar telegraph poles are said to remain floating in the water of the Great Lakes sometimes for several years before they are set in lines and to last better than freshly cut poles.

It is very probable that immersion for long periods in water does materially hasten subsequent seasoning. The tannins, resins, albuminous materials, etc., which are deposited in the cell walls of the fibres of green wood, and which prevent rapid evaporation of the water, undergo changes when under water, probably due to the action of bacteria which live without air, and in the course of time many of these substances are leached out of the wood. The cells thereby become more and more permeable to water, and when the wood is finally brought into the air the water escapes very rapidly and very evenly. Herzenstein's statement that wood prepared by immersion and subsequent drying will absorb more preservative, and that with greater rapidity, is certainly borne out by experience in the United States.

It is sometimes claimed that all seasoning preparatory to treatment with a substance like tar oil might be done away with by putting the green wood into a cylinder with the oil and heating to 225 degrees Fahrenheit, thus driving [154] the water off in the form of steam, after which the tar oil would readily penetrate into the wood. This is the basis of the so-called "Curtiss process" of timber treatment. Without going into any discussion of this method of creosoting, it may be said that the same objection made for steaming holds here. In order to get a temperature of 212 degrees Fahrenheit in the center of the treated wood, the outside temperature would have to be raised so high that the strength of the wood might be seriously injured.

A company on the Pacific coast which treats red fir piling asserts that it avoids this danger by leaving the green timber in the tar oil at a temperature which never exceeds 225 degrees Fahrenheit for from five to twelve hours, until there is no further evidence of water vapor coming out of the wood. The tar oil is then run out, and a vacuum is created for about an hour, after which the oil is run in again and is kept in the cylinders under 100 pounds pressure for from ten to twelve hours, until the required amount of absorption has been reached (about 12 pounds per cubic foot).

Out-of-door Seasoning

The most effective seasoning is without doubt that obtained by the uniform, slow drying which takes place in properly constructed piles outdoors, under exposure to the winds and the sun. Lumber has always been seasoned in this way, which is still the best for ordinary purposes.

It is probable for the sake of economy, air-drying will be eliminated in the drying process of the future without loss to the quality of the product, but as yet no effective method has been discovered whereby this may be accomplished, because nature performs certain functions in air-drying that cannot be duplicated by artificial means. Because of this, hardwoods, as a rule, cannot be successfully kiln-dried green or direct from the saw, and must receive a certain amount of preliminary air-drying before being placed in a dry kiln.

The present methods of air-seasoning in use have been determined by long experience, and are probably as good [155] as they could be made for present conditions. But the same care has not up to this time been given to the seasoning of such timber as ties, bridge material, posts, telegraph and telephone poles, etc. These have sometimes been piled more or less intelligently, but in the majority of cases their value has been too low to make it seem worth while to pile with reference to anything beyond convenience in handling.

In piling material for air-seasoning, one should utilize high, dry ground when possible, and see that the foundations are high enough off the ground, so that there is proper air circulation through the bottom of the piles, and also that the piles are far enough apart so that the air may circulate freely through and around them.

It is air circulation that is desired in all cases of drying, both in dry kilns and out-of-doors, and not sunshine; that is, not the sun shining directly upon the material. The ends also should be protected from the sun, and everything possible done to induce a free circulation of air, and to keep the foundations free from all plant growth.

Naturally, the heavier the material to be dried, the more difficulty is experienced from checking, which has its most active time in the spring when the sap is rising. In fact the main period of danger in material checking comes with the March winds and the April showers, and not infrequently in the South it occurs earlier than that. In other words, as soon as the sap begins to rise, the timber shows signs of checking, and that is the time to take extra precautions by careful piling and protection from the sun. When the hot days of summer arrive the tendency to check is not so bad, but stock will sour from the heat, stain from the sap, mildew from moisture, and fall a prey to wood-destroying insects.

It has been proven in a general way that wood will season more slowly in winter than in summer, and also that the water content during various months varies. In the spring the drying-out of wood cut in October and November will take place more rapidly.

SECTION XI [156]

KILN-DRYING OF WOOD

Advantages of Kiln-drying over Air-drying

Some of the advantages of kiln-drying to be secured over air-drying in addition to reducing the shipping weight and lessening quantity of stock are the following:

- 1. Less material lost.
- 2. Better quality of product.
- 3. Prevention of sap stain and mould.
- 4. Fixation of gums and resins.
- 5. Reduction of hygroscopicity.

This reduction in the tendency to take up moisture means a reduction in the "working" of the material which, even though slight, is of importance.

The problem of drying wood in the best manner divides itself into two distinct parts, one of which is entirely concerned with the behavior of the wood itself and the physical phenomena involved, while the other part has to do with the control of the drying process.

Physical Conditions governing the Drying of Wood

1. Wood is soft and plastic while hot and moist, and becomes "set" in whatever shape it dries. Some species are much more plastic than others.

2. Wood substance begins to shrink only when it dries below the fibre-saturation point, at which it contains from 25 to 30 per cent moisture based on its dry weight. Eucalyptus and certain other species appear to be exceptions to this law.

3. The shrinkage of wood is about twice as great circumferentially as in the radial direction; lengthwise, it is very slight.

4. Wood shrinks most when subjected, while kept moist, to slow drying at high temperatures. [157]

5. Rapid drying produces less shrinkage than slow drying at high temperatures, but is apt to cause case-hardening and honeycombing, especially in dense woods.

6. Case-hardening, honeycombing, and cupping result directly from conditions 1, 4, and 5, and chemical changes of the outer surface.

7. Brittleness is caused by carrying the drying process too far, or by using too high temperatures. Safe limits of treatment vary greatly for different species.

8. Wood absorbs or loses moisture in proportion to the relative humidity in the air, not according to the temperature. This property is called its "hygroscopicity."

9. Hygroscopicity and "working" are reduced but not eliminated by thorough drying.

10. Moisture tends to transfuse from the hot towards the cold portion of the wood.

11. Collapse of the cells may occur in some species while the wood is hot and plastic. This collapse is independent of subsequent shrinkage.

Theory of Kiln-drying

The dry kiln has long since acquired particular appreciation at the hands of those who have witnessed its time-saving qualities, when practically applied to the drying of timber. The science of drying is itself of the simplest, the exposure to the air being, indeed, the only means needed where the matter of time is not called into question. Otherwise, where hours, even minutes, have a marked significance, then other means must be introduced to bring about the desired effect. In any event, however, the same simple and natural remedy pertains,—the absorption of moisture. This moisture in green timber is known as "sap", which is itself composed of a number of ingredients, most important among which are water, resin, and albumen.

All dry kilns in existence use heat to season timber; [158] that is, to drive out that portion of the "sap" which is volatile.

The heat does not drive out the resin of the pines nor the albumen of the hardwoods. It is really of no advantage in this respect. Resin in its hardened state as produced by heat is only slowly soluble in water and contains a large proportion of carbon, the most stable form of matter. Therefore, its retention in the pores of the wood is a positive advantage.

To produce the ideal effect the drying must commence at the heart of the piece and work outward, the moisture being removed from the surface as fast as it exudes from the pores of the wood. To successfully accomplish this, adjustments must be available to regulate the temperature, circulation, and humidity according to the variations of the atmospheric conditions, the kind and condition of the material to be dried.

This ideal effect is only attained by the use of a type of dry kiln in which the surface of the lumber is kept soft, the pores being left open until all the moisture within has been volatilized by the heat and carried off by a free circulation of air. When the moisture has been removed from the pores, the surface is dried without closing the pores, resulting in timber that is clean, soft, bright, straight, and absolutely free from stains, checks, or other imperfections.

Now, no matter how the method of drying may be applied, it must be remembered that vapor exists in the atmosphere at all times, its volume being regulated by the capacity of the temperature absorbed. To kiln-dry properly, a free current of air must be maintained, of sufficient volume to carry off this moisture. Now, the capacity of this air for drying depends entirely upon the ability of its temperature to absorb or carry off a larger proportion of moisture than that apportioned by natural means. Thus, it will be seen, a cubic foot of air at 32 degrees Fahrenheit is capable of absorbing only two grains of water, while at 160 degrees, it will dispose of ninety grains. The air, therefore, should be made as dry as possible and caused to move freely, so as to remove all moisture from the surface of the wood as soon as it appears. [159] Thus the heat effects a double purpose, not only increasing the rate of evaporation, but also the capacity of the air for absorption. Where these means are applied, which rely on the heat alone to accomplish this purpose, only that of the moisture which is volatile succumbs, while the albumen and resin becoming hardened under the treatment close up the pores of the wood. This latter result is oft-times accomplished while moisture yet remains and which in an enforced effort to escape bursts open the cells in which it has been confined and creates what is known as "checks."

Therefore, taking the above facts into consideration, the essentials for the successful kiln-drying of wood may be enumerated as follows:

1. The evaporation from the surface of a stick should not exceed the rate at which the moisture transfuses from the interior to the surface.

2. Drying should proceed uniformly at all points, otherwise extra stresses are set up in the wood, causing warping, etc.

3. Heat should penetrate to the interior of the piece before drying begins.

4. The humidity should be suited to the condition of the wood at the start and reduced in the proper ratio as drying progresses. With wet or green wood it should usually be held uniform at a degree which will prevent the surface from drying below its saturation point until all the free water has evaporated, then gradually reduced to remove the hygroscopic moisture.

5. The temperature should be uniform and as high as the species under treatment will stand without excessive shrinkage, collapse, or checking.

6. Rate of drying should be controlled by the amount of humidity in the air and not by the rate of circulation, which should be made ample at all times.

7. In drying refractory hardwoods, such as oak, best results are obtained at a comparatively low temperature. [160] In more easily dried hardwoods, such as maple, and some of the more difficult softwoods, as cypress, the process may be hastened by a higher temperature but not above the boiling point. In many of the softwoods, the rate of drying may be very greatly increased by heating above the boiling point with a large circulation of vapor at atmospheric pressure.

8. Unequal shrinkage between the exterior and interior portions of the wood and also unequal chemical changes must be guarded against by temperatures and humidities suited to the species in question to prevent subsequent cupping and warping.

9. The degree of dryness attained should conform to the use to which the wood is put.

10. Proper piling of the material and weighting to prevent warping are of great importance.

Requirements in a Satisfactory Dry Kiln

The requirements in a satisfactory dry kiln are:

- 1. Control of humidity at all times.
- 2. Ample air circulation at all points.
- 3. Uniform and proper temperatures.

In order to meet these requirements the United States Forestry Service has designed a kiln in which the humidity, temperature, and circulation can be controlled at all times.

Briefly, it consists of a drying chamber with a partition on either side, making two narrow side chambers open top and bottom.

The steam pipes are in the usual position underneath the material to be dried.

At the top of the side chambers is a spray; at the bottom are gutters and an eliminator or set of baffle plates to separate the fine mist from the air.

The spray accomplishes two things: It induces an increased circulation and it regulates the humidity. This is done by regulating the temperature of the spray water.

The air under the heating coil is saturated at whatever [161] temperature is required. This temperature is the dew point of the air after it passes up into the drying chamber above the coils. Knowing the temperature in the drying room and the dew point, the relative humidity is thus determined.

The relative humidity is simply the ratio of the vapor pressure at the dew point to the pressure of saturated vapor (see Fig. 30).

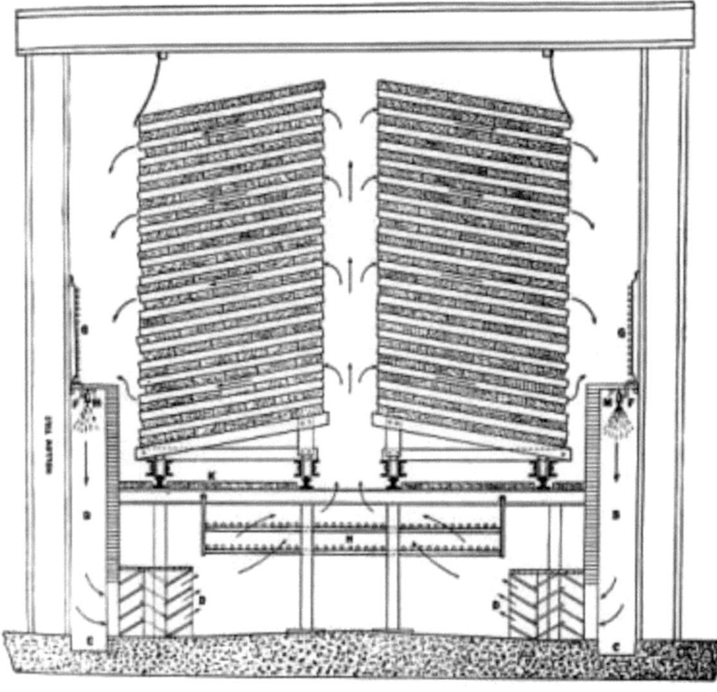

Fig. 30. Section through United States Forestry Service Humidity-controlled Dry Kiln.

Theory and Description of the Forestry Service Kiln

The humidities and temperatures in the piles of lumber are largely dependent upon the circulation of air within the kiln. The temperature and humidity within the kiln, taken alone, are no criterion of the conditions of drying the pile of lumber if the circulation in any portion [162] is deficient. It is possible to have an extremely rapid circulation of air within the dry kiln itself and yet have stagnation within the individual piles, the air passing chiefly through open spaces and channels. Wherever stagnation exists or the movement of air is too sluggish the temperature will drop and the humidity increase, perhaps to the point of saturation.

When in large kilns the forced circulation is in the opposite direction from that induced by the cooling of the air by the lumber, there is always more or less uncertainty as to the movement of the air through the piles. Even with the boards placed edge-wise, with stickers running vertically, and with the heating pipes beneath the lumber, it was found that although the air passed upward through most of the spaces it was actually descending through others, so that very unequal drying resulted. While edge piling would at first thought seem ideal for the freest circulation in an ordinary kiln with steam pipes below, it in fact produces an indeterminate condition; air columns may pass downward through some channels as well as upward through others, and probably stagnate in still others. Nevertheless, edge piling is greatly superior to flat piling where the heating system is below the lumber.

From experiments and from study of conditions in commercial kilns the idea was developed of so arranging the parts of the kiln and the pile of lumber that advantage might be taken of this cooling of the air to assist the circulation. That this can be readily accomplished without doing away with the present features of regulation of humidity by means of a spray of water is clear from Fig. 30, which shows a cross-section of the improved humidity-regulated dry kiln.

In the form shown in the sketch a chamber or flue B runs through the center near the bottom. This flue is only about 6 or 7 feet in height and, together with the water spray F and the baffle plates DD, constitutes the humidity-control feature of the kiln. This control of humidity is affected by the temperature of the water used in the spray. This spray completely saturates the air in the flue B at whatever predetermined temperature [163] is required. The baffle plates DD are to separate all entrained particles of water from the air, so that it is delivered to the heaters in a saturated condition at the required temperature. This temperature is, therefore, the dew point of the air when heated above, and the method of humidity control may therefore be called the dew-point method. It is a very simple matter by means of the humidity diagram (see Fig. 93), or by a hygrodeik (Fig. 94), to determine what dew-point temperature is needed for any desired humidity above the heaters.

Besides regulating the humidity the spray F also acts as an ejector and forces circulation of air through the flue B. The heating system H is concentrated near the outer walls, so as to heat the rising column of air. The temperature within the drying chamber is controlled by means of any suitable thermostat, actuating a valve on the main steam line. The lumber is piled in such a way that the stickers slope downward toward the sides of the kiln.

M is an auxiliary steam spray pointing downward for use at very high temperatures. C is a gutter to catch the precipitation and conduct it back to the pump, the water being recirculated through the sprays. G is a pipe condenser for use toward the end of the drying operation. K is a baffle plate for diverting the heated air and at the same time shielding the under layers of boards from direct radiation of the steam pipes.

The operation of the kiln is simple. The heated air rises above the pipes HH and between the piles of lumber. As it comes in contact with the piles, portions of it are cooled and pass downward and outward through the layers of boards into the space between the condensers GG. Here the column of cooled air descends into the spray flue B, where its velocity is increased by the force of the water spray. It then passes out from the baffle plates to the heaters and repeats the cycle.

One of the greatest advantages of this natural circulation method is that the colder the lumber when placed in the kiln the greater is the movement produced, under the very conditions which call for the greatest circulation—just the opposite of the direct-circulation method. This [164] is a feature of the greatest importance in winter, when the lumber is put into the kiln in a frozen condition. One truckload of lumber at 60 per cent moisture may easily contain over 7,000 pounds of ice.

In the matter of circulation the kiln is, in fact, seldom regulatory—the colder the lumber the greater the circulation produced, with the effect increased toward the cooler and wetter portions of the pile.

Preliminary steaming may be used in connection with this kiln, but experiments indicate that ordinarily it is not desirable, since the high humidity which can be secured gives as good results, and being at as low a temperature as desired, much better results in the case of certain difficult woods like oak, eucalyptus, etc., are obtained.

This kiln has another advantage in that its operation is entirely independent of outdoor atmospheric conditions, except that barometric pressure will effect it slightly.

KILN-DRYING

Remarks

Drying is an essential part of the preparation of wood for manufacture. For a long time the only drying process used or known was air-drying, or the exposure of wood to the gradual drying influences of the open air, and is what has now been termed "preliminary seasoning." This method is without doubt the most successful and effective seasoning, because nature performs certain functions in air-drying that cannot be duplicated by artificial means. Because of this, hardwoods, as a rule, cannot be successfully kiln-dried green or direct from the saw.

Within recent years, considerable interest is awakening among wood users in the operation of kiln-drying. The losses occasioned in air-drying and in improper kiln-drying, and the necessity for getting material dry as quickly as possible from the saw, for shipping purposes and also for manufacturing, are bringing about a realization of the importance of a technical knowledge of the subject. [165]

The losses which occur in air-drying wood, through checking, warping, staining, and rotting, are often greater than one would suppose. While correct statistics of this nature are difficult to obtain, some idea may be had of the amount of degrading of the better class of lumber. In the case of one species of soft wood, Western larch, it is commonly admitted that the best grades fall off sixty to seventy per cent in air-drying, and it is probable that the same is true in the case of Southern swamp oaks. In Western yellow pine, the loss is great, and in the Southern red gum, it is probably as much as thirty per cent. It may be said that in all species there is some loss in air-drying, but in some easily dried species such as spruce, hemlock, maple, etc., it is not so great.

It would hardly be correct to state at the present time that this loss could be entirely prevented by proper methods of kiln-drying the green lumber, but it is safe to say that it can be greatly reduced.

It is well where stock is kiln-dried direct from the saw or knife, after having first been steamed or boiled — as in the case of veneers, etc., — to get them into the kiln while they are still warm, as they are then in good condition for kiln-drying, as the fibres of the wood are soft and the pores well opened, which will allow of forcing the evaporation of moisture without much damage being done to the material.

With softwoods it is a common practice to kiln-dry direct from the saw. This procedure, however, is ill adapted for the hardwoods, in which it would produce such warping and checking as would greatly reduce the value of the product. Therefore, hardwoods, as a rule, are more or less thoroughly air-dried before being placed in the dry kiln, where the residue of moisture may be reduced to within three or four per cent, which is much lower than is possible by air-drying only.

It is probable that for the sake of economy, air-drying will be eliminated in the drying processes of the future without loss to the quality of the product, but as yet no method has been discovered whereby this may be accomplished.

The dry kiln has been, and probably still is, one of the [166] most troublesome factors arising from the development of the timber industry. In the earlier days, before power machinery for the working-up of timber products came into general use, dry kilns were unheard-of, air-drying or seasoning was then relied upon solely to furnish the craftsman with dry stock from which to manufacture his product. Even after machinery had made rapid and startling strides on its way to perfection, the dry kiln remained practically an unknown quantity, but gradually, as the industry developed and demand for dry material increased, the necessity for some more rapid and positive method of seasoning became apparent, and the subject of artificial drying began to receive the serious attention of the more progressive and energetic members of the craft.

Kiln-drying which is an artificial method, originated in the effort to improve or shorten the process, by subjecting the wood to a high temperature or to a draught of heated air in a confined space or kiln. In so doing, time is saved and a certain degree of control over the drying operation is secured.

The first efforts in the way of artificial drying were confined to aiding or hastening nature in the seasoning process by exposing the material to the direct heat from fires built in pits, over which the lumber was piled in a way to expose it to the heat rays of the fires below. This, of course, was a primitive, hazardous, and very unsatisfactory method, to say the least, but it marked the first step in the evolution of the present-day dry kiln, and in that particular only is it deserving of mention.

Underlying Principles

In addition to marking the first step in artificial drying, it illustrated also, in the simplest manner possible, the three underlying principles governing all drying problems: (1)

The application of heat to evaporate or volatilize the water contained in the material; (2) with sufficient air in circulation to carry away in suspension the vapor thus liberated; and (3) with a certain amount of humidity present to prevent the surface from drying too rapidly while the heat is allowed to penetrate to the interior. The [167] last performs two distinct functions: (a) It makes the wood more permeable to the passage of the moisture from the interior of the wood to the surface, and (b) it supplies the latent heat necessary to evaporate the moisture after it reaches the surface. The air circulation is important in removing the moisture after it has been evaporated by the heat, and ventilation also serves the purpose of bringing the heat in contact with the wood. If, however, plain, dry heat is applied to the wood, the surface will become entirely dry before the interior moisture is even heated, let alone removed. This condition causes "case-hardening" or "hollow-horning." So it is very essential that sufficient humidity be maintained to prevent the surface from drying too rapidly, while the heat is allowed to penetrate to the interior.

This humidity or moisture is originated by the evaporation from the drying wood, or by the admission of steam into the dry kiln by the use of steam spray pipes, and is absolutely necessary in the process of hastening the drying of wood. With green lumber it keeps the sap near the surface of the piece in a condition that allows the escape of the moisture from its interior; or, in other words, it prevents the outside from drying first, which would close the pores and cause case-hardening.

The great amount of latent heat necessary to evaporate the water after it has reached the surface is shown by the fact that the evaporation of only one pound of water will extract approximately 66 degrees from 1,000 cubic feet of air, allowing the air to drop in temperature from 154 to 84 degrees Fahrenheit. In addition to this amount of heat, the wood and the water must also be raised to the temperature at which the drying is to be accomplished.

It matters not what type of dry kiln is used, source or application of heating medium, these underlying principles remain the same, and must be the first things considered in the design or selection of the equipment necessary for producing the three essentials of drying: Heat, humidity, and circulation.

Although these principles constitute the basis of all drying problems and must, therefore, be continually [168] carried in mind in the consideration of them, it is equally necessary to have a comprehensive understanding of the characteristics of the materials to be dried, and its action during the drying process. All failures in the past, in the drying of timber products, can be directly attributed to either the kiln designer's neglect of these things, or his failure to carry them fully in mind in the consideration of his problems.

Wood has characteristics very much different from those of other materials, and what little knowledge we have of it and its properties has been taken from the accumulated records of experience. The reason for this imperfect knowledge lies in the fact that wood is not a homogeneous material like the metals, but a complicated structure, and so variable that one stick will behave in a manner widely different from that of another, although it may have been cut from the same tree.

The great variety of woods often makes the mere distinction of the kind or species of the tree most difficult. It is not uncommon to find men of long experience disagree as to the kind of tree a certain piece of lumber was cut from, and, in some cases, there is even a wide difference in the appearance and evidently the structure of timber cut from the same tree.

Objects of Kiln-drying

The objects of kiln-drying wood may be placed under three main headings: (1) To reduce shipping expenses; (2) to reduce the quantity necessary to maintain in stock; and (3) to reduce losses in air-drying and to properly prepare the wood for subsequent use. Item number 2 naturally follows as a consequence of either 1 or 3. The reduction in weight on account of shipping expenses is of greatest significance with the Northwestern lumbermen in the case of Douglas fir, redwood, Western red cedar, sugar pine, bull pine, and other softwoods.

Very rapid methods of rough drying are possible with some of these species, and are in use. High temperatures are used, and the water is sometimes boiled off from the wood by heating above 212 degrees Fahrenheit. These [169] high-temperature methods will not apply to the majority of hardwoods, however, nor to many of the softwoods.

It must first of all be recognized that the drying of lumber is a totally different operation from the drying of a fabric or of thin material. In the latter, it is largely a matter of evaporated moisture, but wood is not only hygroscopic and attracts moisture from the air, but its physical behavior is very complex and renders the extraction of moisture a very complicated process.

An idea of its complexity may be had by mentioning some of the conditions which must be contended with. Shrinkage is, perhaps, the most important. This is unequal in different directions, being twice as great tangentially as radially and fifty times as great radially as longitudinally. Moreover, shrinkage is often unequal in different portions of the same piece. The slowness of the transfusion of moisture through the wood is an important factor. This varies with different woods and greatly in different directions. Wood becomes soft and plastic when hot and moist, and will yield more or less to internal stresses. As some species are practically impervious to air when wet, this plasticity of the cell walls causes them to collapse as the water passes outward from the cell cavities. This difficulty has given much trouble in the case of Western red cedar, and also to some extent in redwood. The unequal shrinkage causes internal stresses in the wood as it dries, which results in warping, checking, case-hardening, and honeycombing.

Case-hardening is one of the most common defects in improperly dried lumber. It is clearly shown by the cupping of the two halves when a case-hardened board is resawed. Chemical changes also occur in the wood in drying, especially so at higher temperatures, rendering it less hygroscopic, but more brittle. If dried too much or at too high a temperature, the strength and toughness is seriously reduced.

Conditions of Success

Commercial success in drying therefore requires that the substance be exposed to the air in the most efficient manner; that the temperature of the air be as high as the [170] substance will stand without injury, and that the air change or movement be as rapid as is consistent with economical installation and operation. Conditions of success therefore require the observance of the following points, which embody the basic principles of the process: (1) The timber should be heated through before drying begins. (2) The air should be very humid at the beginning of the drying process, and be made drier only gradually. (3) The temperature of the lumber must be maintained uniformly throughout the entire pile. (4) Control of the drying process at any given temperature must be secured by controlling the relative humidity, not by decreasing the circulation. (5) In general, high temperatures permit more rapid drying than do lower temperatures. The higher the temperature of the lumber, the more efficient is the kiln. It is believed that temperatures as high as the boiling point are not injurious to most woods, providing all other fundamentally important features are taken care of. Some species, however, are not able to stand as high temperatures as others, and (6) the degree of dryness attained, where strength is the prime requisite, should not exceed that at which the wood is to be used.

Different Treatment according to Kind

The rapidity with which water may be evaporated, that is, the rate of drying, depends on the size and shape of the piece and on the structure of the wood. Thin stock can be dried much faster than thick, under the same conditions of temperature, circulation, and humidity. Pine can be dried, as a general thing, in about one third of the time that would be required for oak of the same thickness, although the former contains the more water of the two. Quarter-sawn oak usually requires half again as long as plain oak. Mahogany requires about the same time as plain oak; ash dries in a little less time, and maple, according to the purpose for which it is intended, may be dried in one fifth the time needed for oak, or may require a slightly longer treatment. For birch, the time required is from one half to two thirds, and for poplar and basswood, from, one fifth to one third that required for oak. [171]

All kinds and thicknesses of lumber cannot be dried at the same time in the same kiln. It is manifest that green and air-dried lumber, dense and porous lumber, all require different treatment. For instance, Southern yellow pine when cut green from the log will stand a very high temperature, say 200 degrees Fahrenheit, and in fact this high temperature is necessary together with a rapid circulation of air in order to neutralize the acidity of the pitch which causes the wood to blue and discolor. This lumber requires to be heated up immediately and to be kept hot throughout the length of the kiln. Hence the kiln must not be of such length as to allow of the air being too much cooled before escaping.

Temperature depends

While it is true that a higher temperature can be carried in the kiln for drying pine and similar woods, this does not altogether account for the great difference in drying time, as experience has taught us that even when both woods are dried in the same kiln, under the same conditions, pine will still dry much faster, proving thereby that the structure of the wood itself affects drying.

The aim of all kiln designers should be to dry in the shortest possible time, without injury to the material. Experience has demonstrated that high temperatures are very effective in evaporating water, regardless of the degree of humidity, but great care must be exercised in using extreme temperatures that the material to be dried is not damaged by checking, case-hardening, or hollow-horning.

The temperature used should depend upon the species and condition of the material when entering the kiln. In general, it is advantageous to have as high a temperature as possible, both for economy of operation and speed of drying, but the physical properties of the wood will govern this.

Many species cannot be dried satisfactorily at high temperatures on account of their peculiar behavior. This is particularly so with green lumber.

Air-dried wood will stand a relatively higher temperature, as a rule, than wet or green wood. In drying green [172] wood direct from the saw, it is usually best to start with a comparatively low temperature, and not raise the temperature until the wood is nearly dry. For example, green maple containing about 60 per cent of its dry weight in water should be started at about 120 degrees Fahrenheit and when it reaches a dryness of 25 per cent, the temperature may be raised gradually up to 190 degrees.

It is exceedingly important that the material be practically at the same temperature throughout if perfect drying is to be secured. It should be the same temperature in the center of a pile or car as on the outside, and the same in the center of each individual piece of wood as on its surface. This is the effect obtained by natural air-drying. The outside atmosphere and breezes (natural air circulation) are so ample that the heat extracted for drying does not appreciably change the temperature.

When once the wood has been raised to a high temperature through and through and especially when the surface has been rendered most permeable to moisture, drying may proceed as rapidly as it can be forced by artificial circulation, provided the heat lost from the wood through vaporization is constantly replaced by the heat of the kiln.

It is evident that to secure an even temperature, a free circulation of air must be brought in contact with the wood. It is also evident that in addition to heat and a circulation of air, the air must be charged with a certain amount of moisture to prevent surface drying or case-hardening.

There are some twenty-five different makes of dry kilns on the market, which fulfill to a varying degree the fundamental requirements. Probably none of them succeed perfectly in fulfilling all.

It is well to have the temperature of a dry kiln controlled by a thermostat which actuates the valve on the main steam supply pipe. It is doubly important to maintain a uniform temperature and avoid fluctuations in the dry kiln, since a change in temperature will greatly alter the relative humidity.

In artificial drying, temperatures of from 150 to 180 degrees [173] Fahrenheit are usually employed. Pine, spruce, cypress, cedar, etc., are dried fresh from the saw, allowing four days for 1-inch stuff. Hardwoods, especially oak, ash, maple, birch, sycamore, etc., are usually air-seasoned for three to six months to allow the first shrinkage to take place more gradually, and are then exposed to the above temperatures in the kiln for about six to ten days for 1-inch stuff, other dimensions in proportion.

Freshly cut poplar and cottonwood are often dried direct from the saw in a kiln. By employing lower temperatures, 100 to 120 degrees Fahrenheit, green oak, ash, etc., can be seasoned in dry kilns without much injury to the material.

Steaming and sweating the wood is sometimes resorted to in order to prevent checking and case-hardening, but not, as has been frequently asserted, to enable the material to dry.

Air Circulation

Air circulation is of the utmost importance, since no drying whatever can take place when it is lacking. The evaporation of moisture requires heat and this must be supplied by the circulating air. Moreover, the moisture laden air must be constantly removed and fresh, drier air substituted. Probably this is the factor which gives more trouble in commercial operations than anything else, and the one which causes the greatest number of failures.

It is necessary that the air circulate through every part of the kiln and that the moving air come in contact with every portion of the material to be dried. In fact, the humidity is dependent upon the circulation. If the air stagnates in any portion of the pile, then the temperature will drop and the humidity rise to a condition of saturation. Drying will not take place at this portion of the pile and the material is apt to mould and rot.

The method of piling the material on trucks or in the kiln, is therefore, of extreme importance. Various methods are in use. Ordinary flat piling is probably the poorest. Flat piling with open chimney spaces in the piles is better. [174] But neither method is suitable for a kiln in which the circulation is mainly vertical.

Edge piling with stickers running vertically is in use in kilns when the heating coils are beneath. This is much better.

Air being cooled as it comes in contact with a pile of material, becomes denser, and consequently tends to sink. Unless the material to be dried is so arranged that the air can pass gradually downward through the pile as it cools, poor circulation is apt to result.

In edge-piled lumber, with the heating system beneath the piles, the natural tendency of the cooled air to descend is opposed by the hot air beneath which tends to rise. An indeterminate condition is thus brought about, resulting in non-uniform drying. It has been found that air will rise through some layers and descend through others.

Humidity

Humidity is of prime importance because the rate of drying and prevention of checking and case-hardening are largely dependent thereon. It is generally true that the surface of the wood should not dry more rapidly than the moisture transfuses from the center of the piece to its surface, otherwise disaster will result. As a sufficient amount of moisture is removed from the wood to maintain the desired humidity, it is not good economy to generate moisture in an outside apparatus and force it into a kiln, unless the moisture in the wood is not sufficient for this purpose; in that case provision should be made for adding any additional moisture that may be required.

The rate of evaporation may best be controlled by controlling the amount of vapor present in the air (relative humidity); it should not be controlled by reducing the air circulation, since a large circulation is needed at all times to supply the necessary heat.

The humidity should be graded from 100 per cent at the receiving end of the kiln, to whatever humidity corresponds with the desired degree of dryness at the delivery end. [175]

The kiln should be so designed that the proper degree may be maintained at its every section.

A fresh piece of sapwood will lose weight in boiling water and can also be dried to quite an extent in steam. This proves conclusively that a high degree of humidity does not have the detrimental effect on drying that is commonly attributed to it. In fact, a proper degree of humidity, especially in the loading or receiving end of a kiln, is just as necessary to good results in drying as getting the proper temperature.

Experiments have demonstrated also that injury to stock in the way of checking, warping, and hollow-horning always develops immediately after the stock is taken into the kiln, and is due to the degree of humidity being too low. The receiving end of the kiln should always be kept moist, where the stock has not been steamed before being put into the kiln. The reason for this is simple enough. When the air is too dry it tends to dry the outside of the material first—which is termed "case-hardening"—and in so doing shrinks and closes up the pores of the wood. As the stock is moved down the kiln, it absorbs a continually increasing amount of heat, which tends to drive off the moisture still present in the center of the stock. The pores on the outside having been closed up, there is no exit for the vapor or steam that is being rapidly formed in the center. It must find its way out some way, and in doing so sets up strains, which result either in checking, warping, or hollow-horning. If the humidity had been kept higher, the outside of the material would not have dried so quickly, and the pores would have remained open for the exit of moisture from the interior of the wood, and this trouble would have been avoided.

Where the humidity is kept at a high point in the receiving end of the kiln, a higher rate of temperature may also be carried, and in that way the drying process is hastened with comparative safety.

It is essential, therefore, to have an ample supply of heat through the convection currents of the air; but in the case of wood the rate of evaporation must be controlled, [176] else checking will occur. This can be done by means of the relative humidity, as stated before. It is clear now that when the air—or, more properly speaking, the space—is completely saturated no evaporation can take place at the given temperature. By reducing the humidity, evaporation takes place more and more rapidly.

Another bad feature of an insufficient and non-uniform supply of heat is that each piece of wood will be heated to the evaporating point on the outer surface, the inside remaining cool until considerable drying has taken place from the surface. Ordinarily in dry kilns high humidity and large circulation of air are antitheses to one another. To obtain the high humidity the circulation is either stopped altogether or greatly reduced, and to reduce the humidity a greater circulation is induced by opening the ventilators or otherwise increasing the draft. This is evidently not good practice, but as a rule is unavoidable in most dry kilns of present make. The humidity should be raised to check evaporation without reducing the circulation if possible.

While thin stock, such as cooperage and box stuff is less inclined to give trouble by undue checking than 1-inch and thicker, one will find that any dry kiln will give more uniform results and, at the same time, be more economical in the use of steam, when the humidity and temperature is carried at as high a point as possible without injury to the material to be dried.

Any well-made dry kiln which will fulfill the conditions required as to circulation and humidity control should work satisfactorily; but each case must be studied by itself, and the various factors modified to suit the peculiar conditions of the problem in hand. In every new case the material should be constantly watched and studied and, if checking begins, the humidity should be increased until it stops. It is not reducing the circulation, but adding the necessary moisture to the air, that should be depended on to prevent checking. For this purpose it is well to have steam jets in the kiln so that if needed they are ready at hand. [177]

Kiln-drying

There are two distinct ways of handling material in dry kilns. One way is to place the load of lumber in a chamber where it remains in the same place throughout the operation, while the conditions of the drying medium are varied as the drying progresses. This is the "apartment" kiln or stationary method. The other is to run the lumber in at one end of the chamber on a wheeled truck and gradually move it along until the drying process is completed, when it is taken out at the opposite end of the kiln. It is the usual custom in these kilns to maintain one end of the chamber moist and the other end dry. This is known as the "progressive" type of kiln, and is the one most commonly used in large operations.

It is, however, the least satisfactory of the two where careful drying is required, since the conditions cannot be so well regulated and the temperatures and humidities are apt to change with any change of wind. The apartment method can be arranged so that it will not require any more kiln space or any more handling of lumber than the progressive type. It does, however, require more intelligent operation, since the conditions in the drying chamber must be changed as the drying progresses. With the progressive type the conditions, once properly established, remain the same.

To obtain draft or circulation three methods are in use — by forced draft or a blower usually placed outside the kiln, by ventilation, and by internal circulation and condensation. A great many patents have been taken out on different methods of ventilation, but in actual operation few kilns work exactly as intended. Frequently the air moves in the reverse direction for which the ventilators were planned. Sometimes a condenser is used in connection with the blower and the air is recirculated. It is also — and more satisfactorily — used with the gentle internal-gravity currents of air.

Many patents have been taken out for heating systems. The differences among these, however, have more to do the mechanical construction than with the process [178] of drying. In general, the heating is either direct or indirect. In the former steam coils are placed in the chamber with the lumber, and in the latter the air is heated by either steam coils or a furnace before it is introduced into the drying chamber.

Moisture is sometimes supplied by means of free steam jets in the kiln or in the entering air; but more often the moisture evaporated from the lumber is relied upon to maintain the humidity necessary.

A substance becomes dry by the evaporation of its inherent moisture into the surrounding space. If this space be confined it soon becomes saturated and the process stops. Hence, constant change is necessary in order that the moisture given off may be continually carried away.

In practice, air movement, is therefore absolutely essential to the process of drying. Heat is merely a useful accessory which serves to decrease the time of drying by increasing both the rate of evaporation and the absorbing power of the surrounding space.

It makes no difference whether this space is a vacuum or filled with air; under either condition it will take up a stated weight of vapor. From this it appears that the vapor molecules find sufficient space between the molecules of air. But the converse is not true, for somewhat less air will be contained in a given space saturated with vapor than in one devoid of moisture. In other words the air does not seem to find sufficient space between the molecules of vapor.

If the temperature of the confined space be increased, opportunity will thereby be provided for the vaporization of more water, but if it be decreased, its capacity for moisture will be reduced and visible water will be deposited. The temperature at which this takes place is known as the "dew-point" and depends upon the initial degree of saturation of the given space; the less the relative saturation the lower the dew-point.

Careful piling of the material to be dried, both in the yard and dry kiln, is essential to good results in drying.

Air-dried material is not dry, and its moisture is too [179] unevenly distributed to insure good behavior after manufacture.

It is quite a difficult matter to give specific or absolute correct weights of any species of timber when thoroughly or properly dried, in order that one may be guided in these kiln operations, as a great deal depends upon the species of wood to be dried, its density, and upon the thickness which it has been cut, and its condition when entering the drying chamber.

Elm will naturally weigh less than beech, and where the wood is close-grained or compact it will weigh more than coarse-grained wood of the same species, and, therefore, no set rules can be laid down, as good judgment only should be used, as the quality of the drying is not purely one of time. Sometimes the comparatively slow process gives excellent results, while to rush a lot of stock through the kiln may be to turn it out so poorly seasoned that it will not give satisfaction when worked into the finished product. The mistreatment of the material in this respect results in numerous defects, chief among which are warping and twisting, checking, case-hardening, and honeycombing, or, as sometimes called, hollow-horning.

Since the proportion of sap and heartwood varies with size, age, species, and individual trees, the following figures as regards weight must be regarded as mere approximations:

Pounds of Water Lost in Drying 100 Pounds of Green Wood in the Kiln

	Sapwood or outer part	Heartwood or interior
(1) Pine, cedar, spruce, and fir	45-65	6-25
(2) Cypress, extremely variable	50-65	18-60
(3) Poplar, cottonwood, and basswood	60-65	40-60
(4) Oak, beech, ash, maple, birch, elm, hickory, chestnut, walnut, and sycamore	40-50	30-40

The lighter kinds have the most water in the sapwood; thus sycamore has more water than hickory, etc.

The efficiency of the drying operations depends a great [180] deal upon the way in which, the lumber is piled, especially when the humidity is not regulated. From the theory of drying it is evident that the rate of evaporation in dry kilns where the humidity is not regulated depends entirely upon the rate of circulation, other things being equal. Consequently, those portions of the wood which receive the greatest amount of air dry the most rapidly, and vice versa. The only way, therefore, in which anything like uniform drying can take place is where the lumber is so piled that each portion of it comes in contact with the same amount of air.

In the Forestry Service kiln (Fig. 30), where the degree of relative humidity is used to control the rate of drying, the amount of circulation makes little difference, provided it exceeds a certain amount. It is desirable to pile the lumber so as to offer as little frictional resistance as possible and at the same time secure uniform circulation. If circulation is excessive in any place it simply means waste of energy but no other injury to the lumber.

The best method of piling is one which permits the heated air to pass through the pile in a somewhat downward direction. The natural tendency of the cooled air to descend is thus taken advantage of in assisting the circulation in the kiln. This is especially important when cold or green lumber is first introduced into the kiln. But even when the lumber has become warmed the cooling due to the evaporation increases the density of the mixture of the air and vapor.

Kiln-drying Gum

The following article was published by the United States Forestry Service as to the best method of kiln-drying gum:

Piling.—Perhaps the most important factor in good kiln-drying, especially in the case of the gums, is the method of piling. It is our opinion that proper and very careful piling will greatly reduce the loss due to warping. A good method of piling is to place the lumber lengthwise of the kiln and on an incline cross-wise. The warm air should rise at [181] the higher side of the pile and descend between the courses of lumber. The reason for this is very simple and the principle has been applied in the manufacture of the best ice boxes for some time. The most efficient refrigerators are iced at the side, the ice compartment opening to the cooling chamber at the top and bottom. The warm air from above is cooled by melting the ice. It then becomes denser and settles down into the main chamber. The articles in the cooling room warm the air as they cool, so it rises to the top and again comes in contact with the ice, thus completing the cycle. The rate of this natural circulation is automatically regulated by the temperature of the articles in the cooling chamber and by the amount of ice in the icing compartment; hence the efficiency of such a box is high.

Now let us apply this principle to the drying of lumber. First we must understand that as long as the lumber is moist and drying, it will always be cooler than the surrounding air, the amount of this difference being determined by the rate of drying and the moisture in the wood. As the lumber dries, its temperature gradually rises until it is equal to that of the air, when perfect dryness results. With this fact in mind it is clear that the function of the lumber in a kiln is exactly analogous to that of the ice in an ice box; that is, it is the cooling agent. Similarly, the heating pipes in a dry kiln bring about the same effect as the articles of food in the ice box in that they serve to heat the air. Therefore, the air will be cooled by the lumber, causing it to pass downward through the piles. If the heating units are placed at the sides of the kiln, the action of the air in a good ice box is duplicated in the kiln. The significant point in this connection is that, the greener and colder the lumber, the faster is the circulation. This is a highly desirable feature.

A second vital point is that as the wood becomes gradually drier the circulation automatically decreases, thus resulting in increased efficiency, because there is no need for circulation greater than enough to maintain the humidity of the air as it leaves the lumber about the same as it enters. Therefore, we advocate either the longitudinal [182] side-wise inclined pile or edge stacking, the latter being much preferable when possible. Of course the piles in our kiln were small and could not be weighted properly, so the best results as to reducing warping were not obtained.

Preliminary Steaming. — Because the fibres of the gums become plastic while moist and hot without causing defects, it is desirable to heat the air-dried lumber to about 200 degrees Fahrenheit in saturated steam at atmospheric pressure in order to reduce the warping. This treatment also furnishes a means of heating the lumber very rapidly. It is probably a good way to stop the sap-staining of green lumber, if it is steamed while green. We have not investigated the other effects of steaming green gum, however, so we hesitate to recommend it.

Temperatures as high as 210 degrees Fahrenheit were used with no apparent harm to the material. The best result was obtained with the temperature of 180 degrees Fahrenheit, after the first preliminary heating in steam to 200 degrees Fahrenheit. Higher temperatures may be used with air-dried gum, however.

The best method of humidity control proved to be to reduce the relative humidity of the air from 100 per cent (saturated steam) very carefully at first and then more rapidly to 30 per cent in about four days. If the change is too marked immediately after the steaming period, checking will invariably result. Under these temperature and humidity conditions the stock was dried from 15 per cent moisture, based on the dry wood weight, to 6 per cent in five days' time. The loss due to checking was about 5 per cent, based on the actual footage loss, not on commercial grades.

Final Steaming. — From time to time during the test runs the material was resawed to test for case-hardening. The stock dried in five days showed slight case-hardening, so it was steamed at atmospheric pressure for 36 minutes near the close of the run, with the result that when dried off again the stresses were no longer present. The material from one run was steamed for three hours at atmospheric [183] pressure and proved very badly case-hardened, but in the reverse direction. It seems possible that by testing for the amount of case-hardening one might select a final steaming period which would eliminate all stresses in the wood.

Kiln-drying of Green Red Gum

The following article was published by the United States Forestry Service on the kiln-drying of green red gum:

A short time ago fifteen fine, red-gum logs 16 feet long were received from Sardis, Miss. They were in excellent condition and quite green.

It has been our belief that if the gum could be kiln-dried directly from the saw, a number of the difficulties in seasoning might be avoided. Therefore, we have undertaken to find out whether or not such a thing is feasible. The green logs now at the laboratory are to be used in this investigation. One run of a preliminary nature has just been made, the method and results of which I will now tell.

This method was really adapted to the drying of Southern pine, and one log of the green gum was cut into 1-inch stock and dried with the pine. The heartwood contained many knots and some checks, although it was in general of quite good quality. The sapwood was in fine condition and almost as white as snow.

This material was edge-stacked with one crosser at either end and one at the center, of the 16-foot board. This is sufficient for the pine, but was absolutely inadequate for drying green gum. A special shrinkage take-up was applied at the three points. The results proved very interesting in spite of the warping which was expected with but three crossers in 16 feet. The method of circulation described was used. It is our belief that edge piling is best for this method.

This method of kiln-drying depends on the maintenance of a high velocity of slightly superheated steam through the lumber. In few words, the object is to maintain the temperature of the vapor as it leaves the lumber at slightly [184] above 212 degrees Fahrenheit. In order to accomplish this result, it is necessary to maintain the high velocity of circulation. As the wood dries, the superheat may be increased until a temperature of 225 degrees or 230 degrees Fahrenheit of the exit air is recorded.

The 1-inch green gum was dried from 20.1 per cent to 11.4 per cent moisture, based on the dry wood weight in 45 hours. The loss due to checking was 10 per cent. Nearly every knot in the heartwood was checked, showing that as the knots could be eliminated in any case, this loss might not be so great. It was significant that practically all of the checking occurred in the heartwood. The loss due to warping was 22 per cent. Of course this was large; but not nearly enough crossers were used for the gum. It is our opinion that this loss due to warping can be very much reduced by using at least eight crossers and providing for taking up of the shrinkage. A feature of this process which is very important is that the method absolutely prevents all sap staining.

Another delightful surprise was the manner in which the superheated steam method of drying changed the color of the sapwood from pure white to a beautifully uniform, clean-looking, cherry red color which very closely resembles that of the heartwood. This method is not new by any means, as several patents have been granted on the steaming of gum to render the sapwood more nearly the color of the heartwoods. The method of application in kiln-drying green gum we believe to be new, however. Other methods for kiln-drying this green stock are to be tested until the proper process is developed. We expect to have something interesting to report in the near future. [1]

[1] The above test was made at the United States Forestry Service Laboratory, Madison, Wis.

SECTION XII [185]

TYPES OF DRY KILNS

DIFFERENT TYPES OF DRY KILNS

Dry kilns as in use to-day are divided into two classes: The "pipe" or "moist-air" kiln, in which natural draft is relied upon for circulation and, the "blower" or "hot blast" kiln, in which the circulation is produced by fans or blowers. Both classes have their adherents and either one will produce satisfactory results if properly operated.

The "Blower" or "Hot Blast" Kiln

The blower kiln in its various types has been in use so long that it is hardly necessary to give to it a lengthy introduction. These kilns at their inauguration were a wonderful improvement over the old style "bake-oven" or "sweat box" kiln then employed, both on account of the improved quality of the material and the rapidity at which it was dried.

These blower kilns have undergone steady improvement, not only in the apparatus and equipment, but also in their general design, method of introducing air, and provision for controlling the temperature and humidity. With this type of kiln the circulation is always under absolute control and can be adjusted to suit the conditions, which necessarily vary with the conditions of the material to be dried and the quantity to be put through the kiln.

In either the blower or moist-air type of dry kiln, however, it is absolutely essential, in order to secure satisfactory results, both as to rapidity in drying and good quality of stock, that the kiln be so designed that the temperature and humidity, together with circulation, are always under convenient control. Any dry kiln in which this has not been carefully considered will not give the desired results. [186]

In the old style blower kiln, while the circulation and temperature was very largely under the operator's control, it was next to impossible to produce conditions in the receiving end of the kiln so that the humidity could be kept at the proper point. In fact, this was one reason why the natural draft, or so-called moist-air kiln was developed.

The advent of the moist-air kiln served as an education to kiln designers and manufacturers, in that it has shown conclusively the value of a proper degree of humidity in the receiving end of any progressive dry kiln, and it has been of special benefit also in that it gave the manufacturers of blower kilns an idea as to how to improve the design of their type of kiln to overcome the difficulty referred to in the old style blower kilns. This has now been remedied, and in a decidedly simple manner, as is usually the case with all things that possess merit.

It was found that by returning from one third to one half of the moist air *after* having passed through the kiln back to the fan room and by mixing it with the fresh and more or less dry air going into the drying room, that the humidity could be kept under convenient control.

The amount of air that can be returned from a kiln of this class depends upon three things: (1) The condition of the material when entering the drying room; (2) the rapidity with which the material is to be dried; and (3) the condition of the outside atmosphere. In the winter season it will be found that a larger proportion of air may be returned to the drying room than in summer, as the air during the winter season contains considerably less moisture and as a consequence is much drier. This is rather a fortunate coincidence, as, when the kiln is being operated in this manner, it will be much more economical in its steam consumption.

In the summer season, when the outside atmosphere is saturated to a much greater extent, it will be found that it is not possible to return as great a quantity of air to the drying room, although there have been instances of kilns of this class, which in operation have had all the air returned and found to give satisfactory results. This is [187] an unusual condition, however, and can only be accounted for by some special or peculiar condition surrounding the installation.

In some instances, the desired amount of humidity in a blower type of kiln is obtained by the addition of a steam spray in the receiving end of the kiln, much in the same manner as that used in the moist-air kilns. This method is not as economical as returning the moisture-laden air from the drying room as explained in the preceding paragraph.

With the positive circulation that may be obtained in a blower kiln, and with the conditions of temperature and humidity under convenient control, this type of kiln has the elements most necessary to produce satisfactory drying in the quickest possible elapsed time.

It must not be inferred from this, however, that this class of dry kiln may be installed and satisfactory results obtained regardless of how it is handled. A great deal of the success of any dry kiln—or any other apparatus, for that matter—depends upon intelligent operation.

Operation of the "Blower" Dry Kiln

It is essential that the operator be supplied with proper facilities to keep a record of the material as it is placed into the drying room, and when it is taken out. An accurate record should be kept of the temperature every two or three hours, for the different thicknesses and species of lumber, that he may have some reliable data to guide him in future cases.

Any man possessing ordinary intelligence can operate dry kilns and secure satisfactory results, providing he will use good judgment and follow the basic instructions as outlined below:

1. When cold and before putting into operation, heat the apparatus slowly until all pipes are hot, then start the fan or blower, gradually bringing it up to its required speed.

2. See that *all* steam supply valves are kept wide open, [188] unless you desire to lengthen the time required to dry the material.

3. When using exhaust steam, the valve from the header (which is a separate drip, independent of the trap connection) must be kept wide open, but must be closed when live steam is used on that part of the heater.

4. The engines as supplied by the manufacturers are constructed to operate the fan or blower at a proper speed with its throttle valve wide open, and with not less than 80 pounds pressure of steam.

5. If the return steam trap does not discharge regularly, it is important that it be opened and thoroughly cleaned and the valve seat re-ground.

6. As good air circulation is as essential as the proper degree of heat, and as the volume of air and its contact with the material to be dried depends upon the volume delivered by the fan or blower, it is necessary to maintain a regular and uniform speed of the engine.

7. Atmospheric openings must always be maintained in the fan or heater room for fresh air supply.

8. Successful drying cannot be accomplished without ample and free circulation of air at all times.

If the above instructions are fully carried out, and good judgment used in the handling and operation of the blower kiln, no difficulties should be encountered in successfully drying the materials at hand.

The "Pipe" or "Moist-air" Dry Kiln

While in the blower class of dry kiln, the circulation is obtained by forced draft with the aid of fans or blowers, in the Moist-air kilns (see Fig. 31); the circulation is obtained by natural draft only, aided by the manipulation of dampers installed at the receiving end of the drying room, which lead to vertical flues through a stack to the outside atmosphere.

The heat in these kilns is obtained by condensing steam in coils of pipe, which are placed underneath the material [189] to be dried. As the degree of heat required, and steam pressure govern the amount of radiation, there are several types of radiating coils. In Fig. 32 will be seen the Single Row Heating Coils for live or high pressure steam, which are used when the low temperature is required. Figure 33 shows the Double (or 2) Row Heating Coils for live or high pressure steam. This apparatus is used when a medium temperature is required. In Fig. 34 will be seen the Vertical Type Heating Coils which is recommended where exhaust or low-pressure steam is to be used, or may be used with live or high-pressure steam when high temperatures are desired.

Fig. 31. Section through a typical Moist-air Dry Kiln.

These heating coils are usually installed in sections, which permit any degree of heat from the minimum to the maximum to be maintained by the elimination of, or the addition of, any number of heating sections. This gives a dry kiln for the drying of green softwoods, or by shutting off a portion of the radiating coils — thus reducing the temperature—a dry kiln for drying hardwoods, that will not stand the maximum degree of heat.

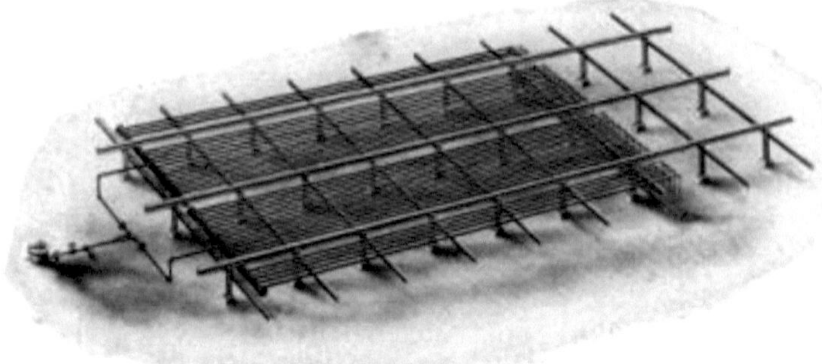

Fig. 32. Single Pipe Heating Apparatus for Dry Kilns, arranged for the Use of Live Steam. For Low Temperatures. [190]

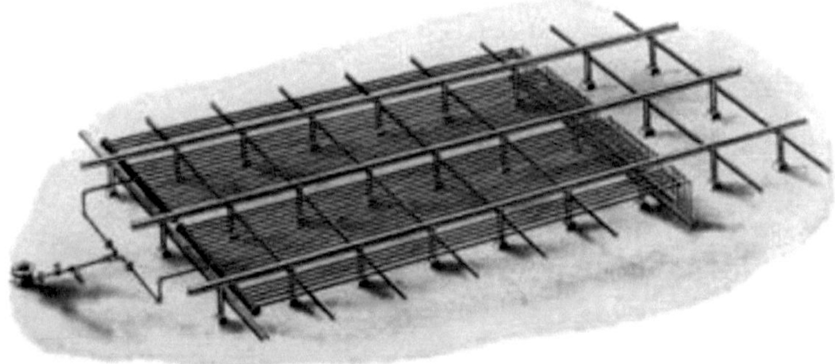

Fig. 33. Double Pipe Heating Apparatus for Dry Kilns, arranged for the Use of Live Steam. For Medium Temperatures. [191]

In the Moist-air or Natural Draft type of dry kiln, any [192] degree of humidity, from clear and dry to a dense fog may be obtained; this is in fact, the main and most important feature of this type of dry kiln, and the most essential one in the drying of hardwoods.

It is not generally understood that the length of a kiln has any effect upon the quantity of material that may be put through it, but it is a fact nevertheless that long kilns are much more effective, and produce a better quality of stock in less time than kilns of shorter length.

Experience has proven that a kiln from 80 to 125 feet in length will produce the best results, and it should be the practice, where possible, to keep them within these figures. The reason for this is that in a long kiln there is a greater drop in temperature between the discharge end and the green or receiving end of the kiln.

It is very essential that the conditions in the receiving end of the kiln, as far as the temperature and humidity are concerned, must go hand in hand.

It has also been found that in a long kiln the desired conditions may be obtained with higher temperatures than with a shorter kiln; consequently higher temperatures may be carried in the discharge end of the kiln, thereby securing greater rapidity in drying. It is not unusual to find that a temperature of 200 degrees Fahrenheit is carried in the discharge end of a long dry kiln with safety, without in any way injuring the quality of the material, although, it would be better not to exceed 180 degrees in the discharge end, and about 120 degrees in the receiving or green end in order to be on the safe side.

Operation of the "Moist-air" Dry Kiln

To obtain the best results these kilns should be kept in continuous operation when once started, that is, they should be operated continuously day and night. When not in operation at night or on Sundays, and the kiln is used to season green stock direct from the saw, the large doors at both ends of the kiln should be opened wide, or the material to be dried will "sap stain."

Fig. 34. Vertical Pipe Heating Apparatus for Dry Kilns; may be used in Connection with either Live or Exhaust Steam for High or Low Temperatures.

It is highly important that the operator attending any drying apparatus keep a minute and accurate record of [193] the condition of the material as it is placed into the drying room, and its final condition when taken out.

Records of the temperature and humidity should be taken frequently and at stated periods for the different thicknesses and species of material, in order that he may have reliable data to guide him in future operations. [194]

The following facts should be taken into consideration when operating the Moist-air dry kiln:

1. Before any material has been placed in the drying room, the steam should be turned into the heating or radiating coils, gradually warming them, and bringing the temperature in the kiln up to the desired degree.

2. Care should be exercised that there is sufficient humidity in the receiving or loading end of the kiln, in order to guard against checking, case-hardening, etc. Therefore it is essential that the steam spray at the receiving or loading end of the kiln be properly manipulated.

3. As the temperature depends principally upon the pressure of steam carried in the boilers, maintain a steam pressure of not less than 80 pounds at all times; it may range as high as 100 pounds. The higher the temperature with its relatively high humidity the more rapidly the drying will be accomplished.

4. Since air circulation is as essential as the proper degree of heat, and as its contact with the material to be dried depends upon its free circulation, it is necessary that the dampers for its admittance into, and its exit from, the drying room be efficiently and properly operated. Successful drying cannot be accomplished without ample and free circulation of air at all times during the drying process.

If the above basic principles are carefully noted and followed out, and good common sense used in the handling and operation of the kiln apparatus, no serious difficulties should arise against the successful drying of the materials at hand. [195]

Choice of Drying Method

At this point naturally arises the question: Which of the two classes of dry kilns, the "Moist-air" or "Blower" kiln is the better adapted for my particular needs?

This must be determined entirely by the species of wood to be dried, its condition when it goes into the kiln, and what kind of finished product is to be manufactured from it.

Almost any species of hardwood which has been subjected to air-seasoning for three months or more may be dried rapidly and in the best possible condition for glue-jointing and fine finishing with a "Blower" kiln, but green hardwood, direct from the saw, can only be successfully dried (if at all) in a "Moist-air" kiln.

Most furniture factories have considerable bent stock which must of necessity be thoroughly steamed before bending. By steaming, the initial process of the Moist-air kiln has been consummated. Hence, the Blower kiln is better adapted to the drying of such stock than the Moist-air kiln would be, as the stock has been thoroughly soaked by the preliminary steaming, and all that is required is sufficient heat to volatilize the moisture, and a strong circulation of air to remove it as it comes to the surface.

The Moist-air kiln is better adapted to the drying of tight cooperage stock, while the Blower kiln is almost universally used throughout the slack cooperage industry for the drying of its products.

For the drying of heavy timbers, planks, blocks, carriage stock, etc., and for all species of hardwood thicker than one inch, the Moist-air kiln is undoubtedly the best.

Both types of kilns are equally well adapted to the drying of 1-inch green Norway and white pine, elm, hemlock, and such woods as are used in the manufacture of flooring, ceiling, siding, shingles, hoops, tub and pail stock, etc.

The selection of one or the other for such work is largely matter of personal opinion. [196]

Kilns of Different Types

All dry kilns as in use to-day are divided as to method of drying into two classes:

- The "Pipe" or "Moist-air" kiln;
- The "Blower" or "Hot Blast" kiln;

both of which have been fully explained in a previous article.

The above two classes are again subdivided into five different types of dry kilns as follows:

- The "Progressive" kiln;
- The "Apartment" kiln;
- The "Pocket" kiln;
- The "Tower" kiln;
- The "Box" kiln.

The "Progressive" Dry Kiln

Dry kilns constructed so that the material goes in at one end and is taken out at the opposite end are called Progressive dry kilns, from the fact that the material gradually progresses through the kiln from one stage to another while drying (see Fig. 31).

In the operation of the Progressive kiln, the material is first subjected to a sweating or steaming process at the receiving or loading end of the kiln with a low temperature and a relative high humidity. It then gradually progresses through the kiln into higher temperatures and lower humidities, as well as changes of air circulation, until it reaches the final stage at the discharge end of the kiln.

Progressive kilns, in order to produce the most satisfactory results, especially in the drying of hardwoods or heavy softwood timbers, should be not less than 100 feet in length (see Fig. 35).

In placing this type of kiln in operation, the following instructions should be carefully followed:

When steam has been turned into the heating coils, and the kiln is fairly warm, place the first car of material to be dried in the drying room — preferably in the morning — about [197] 25 feet from the kiln door on the receiving or loading end of the kiln, blocking the wheels so that it will remain stationary.

Fig. 35. Exterior View of Four Progressive Dry Kilns, each 140 Feet long by 18 Feet wide. Cross-wise piling, fire-proof construction.

Five hours later, or about noon, run in the second car and stop it about five feet from the first one placed in the drying room. Five hours later, or in the evening push car number two up against the first car; then run in car number three, stopping it about five feet from car number two.

On the morning of the second day, push car number three against the others, and then move them all forward about 25 feet, and then run in car number four, stopping it about five feet from the car in advance of it. Five hours later, or about noon, run in car number five and stop it about five feet from car number four. In the evening or about five hours later, push these cars against the ones ahead, and run in loaded car number six, stopping it about five feet from the preceding car.

On the morning of the third day, move all the cars forward about six feet; then run in loaded car number seven stop it about four feet from the car preceding it. [198] Five hours later or about noon push this car against those in advance of it, and run in loaded car number eight moving all cars forward about six feet, and continue in this manner until the full complement of cars have been placed in the kiln. When the kiln has been filled, remove car number one and push all the remaining cars forward and run in the next loaded car, and continue in this manner as long as the kiln is in operation.

As the temperature depends principally upon the pressure of steam, maintain a steam pressure of not less than 80 pounds at all times; it may range up to as high as 100 pounds. The higher the temperature with a relatively higher humidity the more rapidly the drying will be accomplished.

If the above instructions are carried out, the temperatures, humidities, and air circulation properly manipulated, there should be complete success in the handling of this type of dry kiln.

The Progressive type of dry kiln is adapted to such lines of manufacture that have large quantities of material to kiln-dry where the species to be dried is of a similiar nature or texture, and does not vary to any great extent in its thickness, such, for instance, as:

- Oak flooring plants;
- Maple flooring plants;
- Cooperage plants;
- Large box plants;
- Furniture factories; etc.

In the selection of this kind of dry kiln, consideration should be given to the question of ground space of sufficient length or dimension to accommodate a kiln of proper length for successful drying.

The "Apartment" Dry Kiln

The Apartment system of dry kilns are primarily designed for the drying of different kinds or sizes of material at the same time, a separate room or apartment being devoted to each species or size when the quantity is sufficient (see Fig. 36). [199]

These kilns are sometimes built single or in batteries of two or more, generally not exceeding 40 or 50 feet in length with doors and platforms at both ends the same as the Progressive kilns; but in operation each kiln is entirely filled at one loading and then closed, and the entire contents dried at one time, then emptied and again recharged.

Any number of apartments may be built, and each apartment may be arranged to handle any number of cars, generally about three or four, or they may be so constructed that the material is piled directly upon the floor of the drying room.

Fig. 36. Exterior View of Six Apartment Dry Kilns, each 10 Feet wide by 52 Feet long, End-wise Piling. They are entirely of fire-proof construction and equipped with double doors (Hussey asbestos outside and canvas inside), and are also equipped with humidity and air control dampers, which may be operated from the outside without opening the kiln doors, which is a very good feature.

When cars are used, it is well to have a transfer car at each end of the kilns, and stub tracks for holding cars of dry material, and for the loading of the unseasoned stock, as in this manner the kilns may be kept in full operation at all times.

In this type of dry kiln the material receives the same [200] treatment and process that it would in a Progressive kiln. The advantages of Apartment kilns is manifest where certain conditions require the drying of numerous kinds as well as thicknesses of material at one and the same time. This method permits of several short drying rooms or apartments so that it is not necessary to mix hardwoods and softwoods, or thick and thin material in the same kiln room.

In these small kilns the circulation is under perfect control, so that the efficiency is equal to that of the more extensive plants, and will readily appeal to manufacturers whose output calls for the prompt and constant seasoning of a large variety of small stock, rather than a large volume of material of uniform size and grade.

Apartment kilns are recommended for industries where conditions require numerous kinds and thicknesses of material to be dried, such as:

- Furniture factories;
- Piano factories;
- Interior woodwork mills;
- Planing mills; etc.

The "Pocket" Dry Kiln

"Pocket" dry kilns (see Fig. 37) are generally built in batteries of several pockets. They have the tracks level and the lumber goes in and out at the same end. Each drying room is entirely filled at one time, the material is dried and then removed and the kiln again recharged.

The length of "Pocket" kilns ranges generally from 14 feet to about 32 feet.

The interior equipment for this type of dry kiln is arranged very similiar to that used in the Apartment kiln. The heating or radiating coils and steam spray jets extend the whole length of the drying room, and are arranged for the use of either live or exhaust steam, as desired.

Inasmuch as Pocket kilns have doors at one end only, this feature eliminates a certain amount of door exposure, which conduces towards economy in operation. [201]

In operating Pocket kilns, woods of different texture and thickness should be separated and placed in different drying rooms, and each kiln adjusted and operated to accommodate the peculiarities of the species and thickness of the material to be dried.

Fig. 37. Exterior View of Five Pocket Dry Kilns, built in Two Batteries with the Front of each Set facing the other, and a Transfer System between. They are also equipped with the asbestos doors.

Naturally, the more complex the conditions of manufacturing wood products in any industry, the more difficult [202] will be the proper drying of same. Pocket kilns, are, therefore, recommended for factories having several different kinds and thicknesses of material to dry in small quantities of each, such as:

- Planing mills;
- Chair factories;
- Furniture factories;
- Sash and door factories; etc.

The "Tower" Dry Kiln

The so-called "Tower" dry kiln (see Fig. 38) is designed for the rapid drying of small stuff in quantities. Although the general form of construction and the capacity of the individual bins or drying rooms may vary, the same essential method of operation is common to all. That is, the material itself, such as wooden novelties, loose staves, and heading for tubs, kits, and pails, for box stuff, kindling wood, etc., is dumped directly into the drying rooms from above, or through the roof, in such quantities as effectually to fill the bin, from which it is finally removed when dry, through the doors at the bottom.

These dry kilns are usually operated as "Blower" kilns, the heating apparatus is generally located in a separate room or building adjacent to the main structure or drying rooms, and arranged so that the hot air discharged through the inlet duct (see illustration) is thoroughly distributed beneath a lattice floor upon which rests the material to be dried. Through this floor the air passes directly upward, between and around the stock, and finally returns to the fan or heating room.

This return air duct is so arranged that by means of dampers, leading from each drying room, the air may be returned in any quantity to the fan room where it is mixed with fresh air and again used. This is one of the main features of economy of the blower system of drying, as by the employment of this return air system, considerable saving may be made in the amount of steam required for drying.

Fig. 38. Exterior and Sectional View of a Battery of Tower Dry Kilns. This is a "Blower" or "Hot Blast" type, and shows the arrangement of the fan blower, engine, etc. This type of dry kin is used principally for the seasoning of small, loose material.

The lattice floors in this type of dry kiln are built on [203] an incline, which arrangement materially lessens the cost, and increases the convenience with which the dried stock may be removed from the bins or drying rooms.

In operation, the material is conveyed in cars or trucks [204] on an overhead trestle—which is inclosed—from which the material to be dried is dumped directly into the drying rooms or bins, through hoppers arranged for that purpose thereby creating considerable saving in the handling of the material to be dried into the kiln. The entire arrangement thus secures the maximum capacity, with a minimum amount of floor space, with the least expense. Of course, the higher these kilns are built, the less relative cost for a given result in the amount of material dried.

In some instances, these kilns are built less in height and up against an embankment so that teamloads of material may be run directly onto the roof of the kilns, and dumped through the hoppers into the drying rooms or bins, thus again reducing to a minimum the cost of this handling.

The return air duct plays an important part in both of these methods of filling, permitting the air to become saturated to the maximum desired, and utilizing much of the heat contained therein, which would otherwise escape to the atmosphere.

The "Tower" kiln is especially adapted to factories of the following class:

- Sawmills;
- Novelty factories;
- Woodenware factories;
- Tub and pail factories; etc.

The "Box" Dry Kiln

The "Box" kiln shown in Figure 39 is an exterior view of a kiln of this type which is 20 feet wide, 19 feet deep, and 14 feet high, which is the size generally used when the space will permit.

Box kilns are used mostly where only a small quantity of material is to be dried. They are not equipped with trucks or cars, the material to be dried being piled upon a [205] raised platform inside the drying room. This arrangement, therefore, makes them of less cost than the other types of dry kilns.

They are particularly adapted to any and all species and size of lumber to be dried in very small quantities.

Fig. 39. Exterior view of the Box Dry Kiln. This particular kiln is 20 feet wide, 19 feet deep and 14 feet high. Box kilns are used mostly where only a small amount of kiln-dried lumber of various sizes is required. They are not equipped with trucks or cars, and therefore cost less to construct than any other type of dry kiln.

In these small kilns the circulation is under perfect control, so that the efficiency is equal to that of the more extensive plants.

These special kilns will readily appeal to manufacturers, whose output calls for the prompt and constant seasoning of a large variety of small stock, rather than a large volume material of uniform size and grade.

SECTION XIII [206]

DRY KILN SPECIALTIES

KILN CARS AND METHOD OF LOAD-ING

Within recent years, the edge-wise piling of lumber (see Figs. 40 and 41), upon kiln cars has met with considerable favor on account of its many advantages over the older method of flat piling. It has been proven that lumber stacked edge-wise dries more uniformly and rapidly, and with practically no warping or twisting of the material, and that it is finally discharged from the dry kiln in a much better and brighter condition. This method of piling also considerably increases the holding and consequent drying capacities of the dry kiln by reason of the increased carrying capacities of the kiln cars, and the shorter period of time required for drying the material.

Fig. 40. Car Loaded with Lumber on its Edges by the Automatic Stacker, to go into the Dry Kiln cross-wise. Equipped with two edge piling kiln trucks. [207]

229

In Figures 42 and 43 are shown different views of the automatic lumber stacker for edge-wise piling of lumber on kiln cars. Many users of automatic stackers report that the grade of their lumber is raised to such an extent that the system would be profitable for this reason alone, not taking into consideration the added saving in time and labor, which to anyone's mind should be the most important item.

Fig. 41. Car Loaded with Lumber on its Edges by the Automatic Stacker, to go into the Dry Kiln end-wise. The bunks on which the lumber rests are channel steel. The end sockets are malleable iron and made for I-beam stakes.

In operation, the lumber is carried to these automatic stackers on transfer chains or chain conveyors, and passes on to the stacker table. When the table is covered with boards, the "lumber" lever is pulled by the operator, which raises a stop, preventing any more lumber leaving the chain conveyor. The "table" lever then operates the friction drive and raises the table filled with the boards to a vertical position. As the table goes up, it raises the latches, which fall into place behind the piling strips that had been previously laid on the table. When the table returns to the lower position, a new set of piling strips are put in place on the table, and the stream of boards which has been accumulating on the conveyor chain are again permitted to flow onto the table. As each layer of lumber is added, the kiln car is forced out against a strong [208] tension. When the car is

loaded, binders are put on over the stakes by means of a
powerful lever arrangement.

Fig. 42. The above illustration shows the construction of
the Automatic Lumber Stacker for edge piling of lumber to
go into the dry kiln end-wise.

Fig. 43. The above illustration shows the construction of the Automatic Lumber Stacker for edge piling of lumber to go into the dry kiln cross-wise. [209]

Fig. 44. The above illustration shows a battery of Three Automatic Lumber Stackers.

Fig. 45. The above illustration shows another battery of Three Automatic Lumber Stackers. [210]

Fig. 46. Cars Loaded with Lumber on its Edges by the
Automatic Lumber Stackers.

After leaving the dry kilns, the loaded car is transferred
to the unstacker (see Fig. 47). Here it is placed on the unstacker car which, by means of a tension device, holds the
load of lumber tight against the vertical frame of the unstacker. The frame of the unstacker is triangular and has a
series of chains. Each chain has two special links with pro-
jecting lugs. The chains all travel in unison. The lug links
engage a layer of boards, sliding the entire layer vertically,
and the boards, one at a time, fall over the top of the unstacker frame onto the inclined table, and from there onto
conveyor chains from which they may be delivered to any
point desired, depending upon the length and direction of
the chain conveyor.

With these unstackers one man can easily unload a kiln
car in twenty minutes or less.

Fig. 47. The Lumber Unstacker Car, used for unloading
cars of Lumber loaded by the Automatic Stacker.

Fig. 48. The Lumber Unstacker Car and Unstacker, used for unloading Lumber loaded by the Automatic Stacker.

The experience of many users prove that these edge stacking machines are not alike. This is important, because there is one feature of edge stacking that must not be overlooked. Unless each layer of boards is forced [211] into place by power and held under a strong pressure, much slack will accumulate in an entire load, and the subsequent handling of the kiln cars, and the effect of the kiln-drying will loosen up the load until there is a tendency for the layers to telescope. And unless the boards are held in place rigidly and with strong pressure they will have a tendency to warp.

Fig. 49. The above illustration shows method of loading kiln cars with [212] veneer on its edges by the use of the Tilting Platform.

A kiln car of edge-stacked lumber, properly piled, is made up of alternate solid sheets of lumber and vertical open-air spaces, so that the hot air and vapors rise naturally and freely through the lumber, drying both sides of the board evenly. The distribution of the heat and moisture being even and uniform, the drying process is naturally quickened, and there is no opportunity or tendency for the lumber to warp.

In Figure 49 will be seen a method of loading kiln cars with veneer on edge by the use of a tilting platform. On the right of the illustration is seen a partially loaded kiln car tilted to an angle of 45 degrees, to facilitate the placing [213] of the veneer on the car. At the left is a completely loaded car ready to enter the dry kiln.

Gum, poplar, and pine veneers are satisfactorily dried in this manner in from 8 to 24 hours.

In Figure 50 will be seen method of piling lumber on the flat, "cross-wise" of the dry kiln when same has three tracks.

Fig. 50. Method of Loading lumber on its Flat, cross-wise of the Dry Kiln when same has Three Tracks.

In Figure 51 will be seen another method of piling lumber on the flat, "cross-wise" of the dry kiln when same has three tracks.

In Figure 52 will be seen method of piling lumber on the flat, "end-wise" of the dry kiln when same has two tracks.

In Figure 53 will be seen another method of piling lumber on the flat, "end-wise" of the dry kiln when same has two tracks.

In Figure 54 will be seen method of piling slack or tight barrel staves "cross-wise" of the kiln when same has three tracks.

In Figure 55 will be seen another method of piling slack or tight barrel staves "cross-wise" of the dry kiln when same has three tracks.

In Figure 56 will be seen method of piling small tub or pail staves "cross-wise" of the dry kiln when same has two tracks.

In Figure 57 will be seen method of piling bundled staves "cross-wise" of the dry kiln when same has two tracks. [214]

Fig. 51. Method of loading Lumber on its Flat, cross-wise of the Dry Kiln when same has Three Tracks.

Fig. 52. Method of loading Lumber on its Flat, end-wise of the Dry Kiln by the Use of the Single-sill or Dolly Truck. [215]

Fig. 53. Method of loading Lumber on its Flat, end-wise of the Dry Kiln by the Use of the Double-sill Truck. [216]

Fig. 54. Method of loading Kiln Car with Tight or Slack Barrel Staves cross-wise of Dry Kiln.

Fig. 55. Method of loading Kiln Car with Tight or Slack Barrel Staves cross-wise of Dry Kiln. [217]

Fig. 56. Method of loading Kiln Car with Tub or Pail
Staves cross-wise of Dry Kiln.

Fig. 57. Method of loading Kiln Car with Bundled Staves cross-wise of Dry Kiln. [218]

In Figure 58 will be seen method of piling shingles "cross-wise" of dry kiln when same has three tracks.

In Figure 59 will be seen another method of piling shingles "cross-wise" of the dry kiln when same has three tracks.

Fig. 58. Method of loading Kiln Car with Shingles cross-wise of Dry Kiln.

In Figure 60 will be seen method of piling shingles "end-
wise" of the dry kiln when same has two tracks.

In Figure 61 will be seen a kiln car designed for handling
short tub or pail staves through a dry kiln.

Fig. 60. Car loaded with 100,000 Shingles. Equipped with
four long end-wise piling trucks and to go into dry kiln
end-wise.

Fig. 61. Kiln Car designed for handling Short Tub or Pail
Staves through a Dry Kiln. [220]

In Figure 62 will be seen a kiln car designed for short
piece stock through a dry kiln.

In Figure 63 will be seen a type of truck designed for the
handling of stave bolts about a stave mill or through a
steam box.

In Figure 64 will be seen another type of truck designed
for the handling of stave bolts about a stave mill or through
a steam box.

In Figure 65 will be seen another type of truck designed
for the handling of stave bolts about a stave mill or through
a steam box.

In Figure 66 will be seen another type of truck designed for the handling of stave bolts about a stave mill or through a steam box.

In Figure 67 will be seen another type of truck designed for the handling of stave bolts about a stave mill or through a steam box.

In Figure 68 will be seen another type of truck designed for the handling of stave bolts about a stave mill or through a steam box.

In Figure 69 will be seen the Regular 3-rail Transfer Car designed for the handling of 2-rail kiln cars which have been loaded "end-wise."

In Figure 70 will be seen another type of Regular 3-rail Transfer Car designed for the handling of 2-rail kiln cars which have been loaded "end-wise."

In Figure 71 will be seen a Specially-designed 4-rail Transfer Car for 2-rail kiln cars which have been built to accommodate extra long material to be dried.

In Figure 72 will be seen the Regular 2-rail Transfer Car designed for the handling of 3-rail kiln cars which have been loaded "cross-wise."

In Figure 73 will be seen another type of Regular 2-rail Transfer Car designed for the handling of 3-rail kiln cars which have been loaded "cross-wise."

In Figure 74 will be seen the Regular 2-rail Underslung type of Transfer Car designed for the handling of 3-rail kiln cars which have been loaded "cross-wise." Two important features in the construction of this transfer car make it extremely easy in its operation. It has extra large wheels, diameter 131/2 inches, and being underslung, the top of its rails are no higher than the other types of transfer cars. Note the relative size of the wheels in the illustration, yet the car is only about 10 inches in height. [221]

Fig. 62. Kiln Car Designed for handling Short Piece Stock
through a Dry Kiln.

Fig. 63. A Stave Bolt Truck. [222]

Fig. 64. A Stave Bolt Truck.

Fig. 65. A Stave Bolt Truck. [223]

Fig. 66. A Stave Bolt Truck.

Fig. 67. A Stave Bolt Truck. [224]

Fig. 68. A Stave Bolt Truck.

Fig. 69. A Regular 3-Rail Transfer Truck. [225]

Fig. 70. A Regular 3-Rail Transfer Truck.

Fig. 71. A Special 4-Rail Transfer Truck.

Fig. 72. A Regular 2-Rail Transfer Truck. [226]

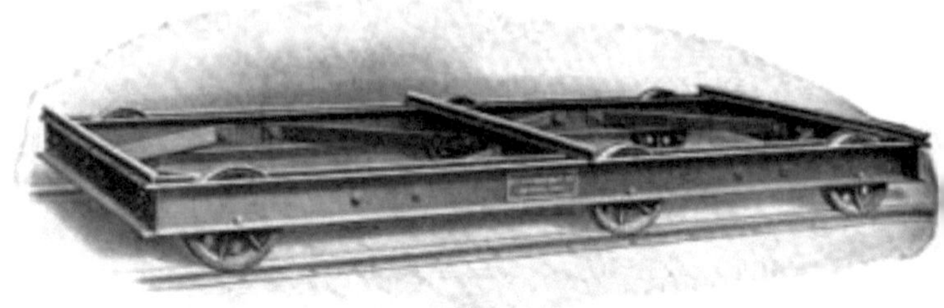

Fig. 73. A Regular 2-Rail Transfer Truck.

Fig. 74. A Regular 2-Rail Underslung Transfer Truck.

In Figure 75 will be seen the Regular 3-rail Underslung type of Transfer Car designed for the handling of 2-rail kiln cars which have been loaded "end-wise." This car also has the important features of large diameter wheels and low rail construction, which make it very easy in its operation.

Fig. 75. A Regular 3-Rail Underslung Transfer Truck.
[227]

In Figure 76 will be seen the Special 2-rail Flexible type of Transfer Car designed for the handling of 3-rail kiln cars which have been loaded "cross-wise." This car is equipped with double the usual number of wheels, and by making each set of trucks a separate unit (the front and rear trucks being bolted to a steel beam with malleable iron connection), a slight up-and-down movement is permitted, which enables this transfer car to adjust itself to any unevenness in the track, which is a very good feature.

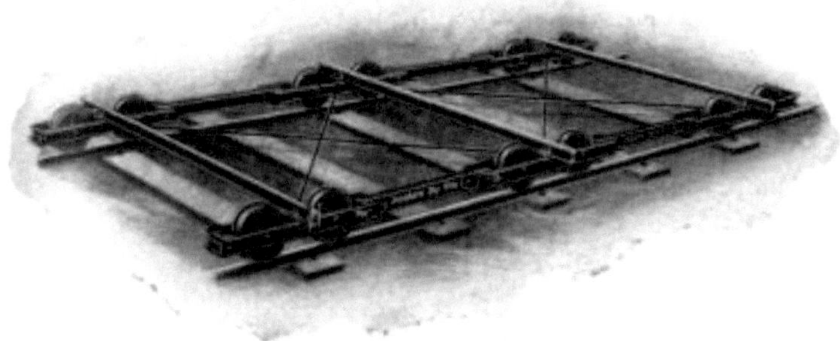

Fig. 76. A Special 2-Rail Flexible Transfer Truck.

In Figure 77 will be seen the Regular Transfer Car designed for the handling of stave bolt trucks.

In Figure 78 will be seen another type of Regular Transfer Car designed for the handling of stave bolt trucks.

In Figure 79 will be seen a Special Transfer Car designed for the handling of stave bolt trucks. [228]

Fig. 77. A Regular Transfer Car for handling Stave Bolt Trucks.

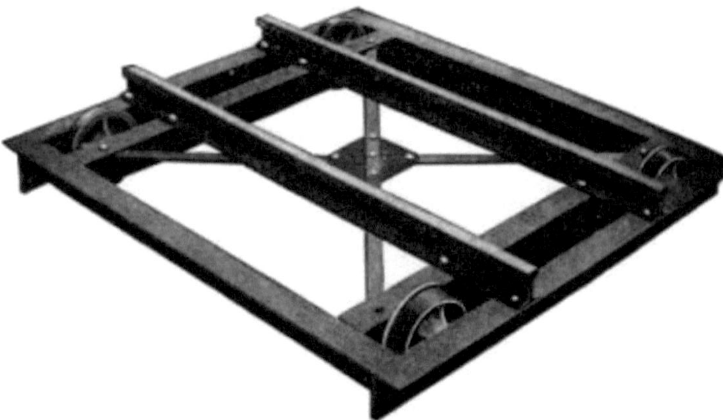

Fig. 78. A Regular Transfer Car for handling Stave Bolt Trucks.

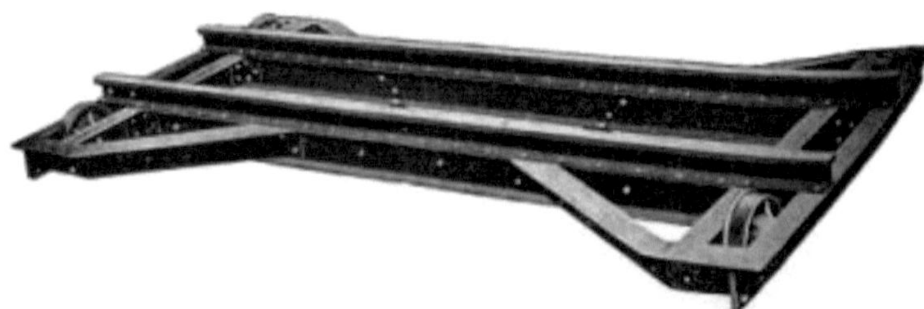

Fig. 79. A Special Transfer Car for handling Stave Bolt Trucks. [229]

In Figure 80 will be seen the Regular Channel-iron Kiln Truck designed for edge piling "cross-wise" of the dry kiln.

In Figure 81 will be seen another type of Regular Channel-iron Kiln Truck designed for edge piling "cross-wise" of the dry kiln.

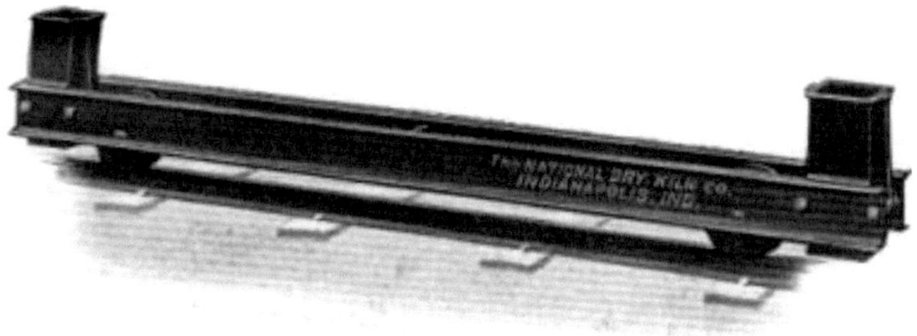

Fig. 80. A Regular Channel-iron Kiln Truck.

Fig. 81. A Regular Channel-iron Kiln Truck. [230]

In Figure 82 will be seen the Regular Channel-iron Kiln Truck designed for flat piling "end-wise" of the dry kiln.

Fig. 82. A Regular Channel-iron Kiln Truck.

In Figure 83 will be seen the Regular Channel-iron Kiln Truck with I-Beam cross-pieces designed for flat piling "end-wise" of the dry kiln.

In Figure 84 will be seen the Regular Small Dolly Kiln Truck designed for flat piling "end-wise" of the dry kiln. [231]

Fig. 83. A Regular Channel-iron Kiln Truck.

Fig. 84. A Regular Single-sill or Dolly Kiln Truck.

Different Types of Kiln Doors

In Figure 85 will be seen the Asbestos-lined Door. The construction of this kiln door is such that it has no tendency to warp or twist. The framework is solid and the body is made of thin slats placed so that the slat on either side covers the open space of the other with asbestos roofing fabric in between. This makes a comparatively light and inexpensive door, and one that absolutely holds the heat. These doors may be built either swinging, hoisting, or sliding.

Fig. 85. An Asbestos-lined Kiln Door of the Hinge Type.

In Figure 86 will be seen the Twin Carrier type of door hangers with doors loaded and rolling clear of the opening. [232]

Fig. 86. Twin Carrier with Kiln Door loaded and rolling clear of Opening.

In Figure 87 will be seen the Twin Carrier for doors 18 to 35 feet wide, idle on a section of the track.

In Figure 88 will be seen another type of carrier for kiln doors.

In Figure 89 will be seen the preceding type of kiln door carrier in operation.

In Figure 90 will be seen another type of carrier for kiln doors.

In Figure 91 will be seen kiln doors seated, wood construction, showing 3½" × 5¾" inch-track timbers and trusses, supported on 4-inch by 6-inch jamb posts. "T" rail track, top and side, inclined shelves on which the kiln door rests. Track timber not trussed on openings under 12 feet wide.

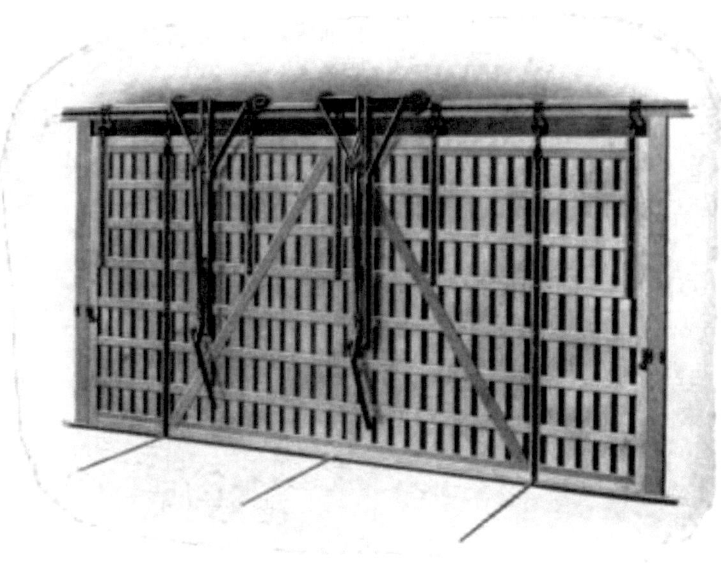

Fig. 87. Twin Carriers for Kiln Doors 18 to 35 Feet wide.
[233]

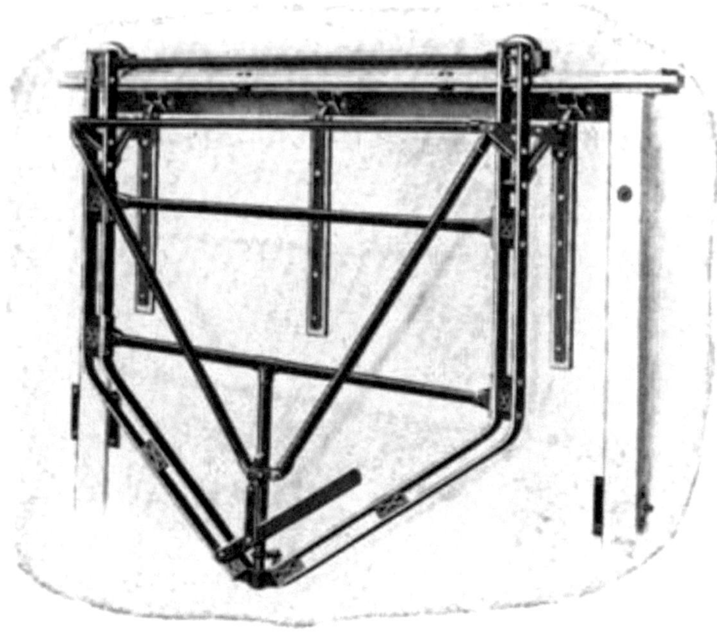

Fig. 88. Kiln Door Carrier engaged to Door Ready for lifting.

In Figure 92 will be seen kiln doors seated, fire-proof
construction, showing 12-inch, channel, steel lintels, 2" × 2"
steel angle mullions, track brackets bolted to the steel lintels and "T" rail track. No track timbers or trusses used.
[234]

Fig. 89. Kiln Door Carrier shown on Doors of Wood Construction. [235]

Fig. 90. Kiln Door Construction with Door Carrier out of Sight.

Fig. 91. Kiln Door Construction. Doors Seated. Wood Construction. [236]

Fig. 92. Kiln Door Construction. Doors Seated. Fire-proof Construction.

SECTION XIV [237]

HELPFUL APPLIANCES IN KILN-DRYING

The Humidity Diagram

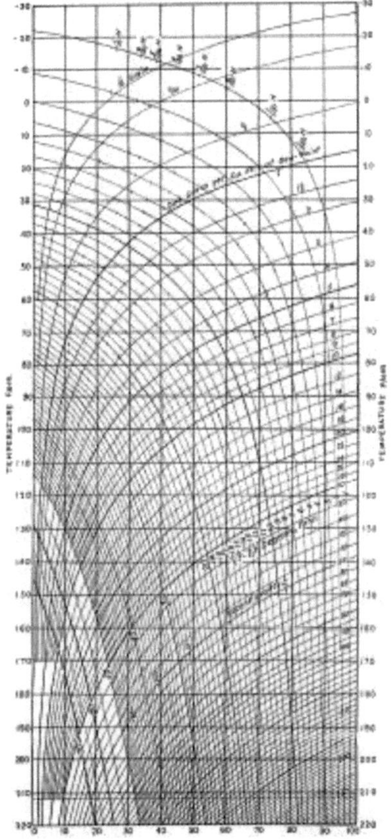

Fig. 93. The United States Forest Service Humidity Diagram for determination of Absolute Humidities. Dew Points and Vapor Pressures; also Relative Humidities by means of Wet and Dry-Bulb Thermometer, for any temperatures and change in temperature.

Some simple means of determining humidities and changes in humidity brought about by changes in temperature in the dry kiln without the use of tables is almost a necessity. To meet this requirement the United States Forestry Service has devised the Humidity Diagram shown in Figure 93. It differs in several respects from the hydrodeiks now in use.

The purpose of the humidity diagram is to enable the dry-kiln operator to determine quickly the humidity conditions and vapor pressure, as well as the changes which take place with changes of temperature. The diagram above is adapted to the direct solution of problems of this character without recourse to tables or mathematical calculations.

The humidity diagram consists of two distinct sets of curves on the same sheet. One set, the convex curves, is for the determination of relative humidity of wet-and-dry-bulb hygrometer or psychrometer; the other, the concave curves, is derived from the vapor pressures and shows the amount of moisture per cubic foot at relative humidities and temperatures when read at the dew-point. The latter curves, therefore, are independent of all variables affecting the wet-bulb readings. They are proportional to vapor pressures, not to density, and, therefore, may be followed from one temperature to another with correctness. The short dashes show the correction (increase or decrease) which is necessary in the relative humidity, read from the convex curves, with an increase or decrease from the normal barometric pressure of 30 inches, for which the curves [238] have been plotted. This correction, except for very low temperatures, is so small that it may usually be disregarded.

The ordinates, or vertical distances, are relative humidity expressed in per cent of saturation, from 0 per cent at the bottom to 100 per cent at the top. The abscissae, or horizontal distances, are temperatures in degrees Fahrenheit from 30 degrees below zero, at the left, to 220 degrees above, at the right.

Examples of Use

The application of the humidity diagram can best be understood by sample problems. These problems also show the wide range of conditions to which the diagram will apply.

Example 1. To find the relative humidity by use of wet-and-dry-bulb hygrometer or psychrometer:

Place the instrument in a strong circulation of air, or wave it to and fro. Read the temperature of the dry bulb and the wet, and subtract. Find on the horizontal line the temperature shown by the dry-bulb thermometer. Follow the vertical line from this point till it intersects with the convex curve marked with the difference between the wet and dry readings. The horizontal line passing through this intersection will give the relative humidity.

Example: Dry bulb 70°, wet bulb 62°, difference 8°. Find 70° on the horizontal line of temperature. Follow up the vertical line from 70° until it intersects with the convex curve marked 8°. The horizontal line passing through this intersection shows the relative humidity to be 64 per cent.

Example 2. To find how much water per cubic foot is contained in the air:

Find the relative humidity as in example 1. Then the nearest concave curve gives the weight of water in grains per cubic foot when the air is cooled to the dew-point. Using the same quantities as in example 1, this will be slightly more than 5 grains.

Example 3. To find the amount of water required to saturate air at a given temperature:

Find on the top line (100 per cent humidity) the given temperature; the concave curve intersecting at or near [239] this point gives the number of grains per cubic foot. (Interpolate, if great accuracy is desired.)

Example 4. To find the dew-point:

Obtain the relative humidity as in example 1. Then follow up parallel to the nearest concave curve until the top horizontal (indicating 100 per cent relative humidity) is reached. The temperature on this horizontal line at the point reached will be the dew-point.

Example: Dry bulb 70°, wet bulb 62°. On the vertical line for 70° find the intersection with the hygrometer (convex) curve for 8°. This will be found at nearly 64 per cent relative humidity. Then follow up parallel with the vapor pressure (concave) curve marked 5 grains to its intersection at the top of the chart with the 100 per cent humidity line. This gives the dew-point as 57°.

Example 5. To find the change in the relative humidity produced by a change in temperature:

Example: The air at 70° Fahr. is found to contain 64 per cent humidity; what will be its relative humidity if heated to 150° Fahr.? Starting from the intersection of the designated humidity and temperature coordinates, follow the vapor-pressure curve (concave) until it intersects the 150° temperature ordinate. The horizontal line then reads 6 per cent relative humidity. The same operation applies to reductions in temperature. In the above example what is the humidity at 60°? Following parallel to the same curve in the opposite direction until it intersects the 60° ordinate gives 90 per cent; at 57° it becomes 100 per cent, reaching the dew-point.

Example 6. To find the amount of condensation produced by lowering the temperature:

Example: At 150° the wet bulb reads 132°. How much water would be condensed if the temperature were lowered to 70°? The intersection of the hygrometer curve for 18° (150°-132°) with temperature line for 150° shows a relative humidity of 60 per cent. The vapor-pressure curve (concave) followed up to the 100 per cent relative humidity line shows 45 grains per cubic foot at the dew-point, which corresponds to a temperature of 130°. At 70° it is seen that the air can contain but 8 grains per cubic foot (saturation).

Consequently, there will be condensed 45 minus 8, or 37 grains per cubic foot of space measured at the dew-point. [240]

Example 7. To find the amount of water required to produce saturation by a given rise in temperature:

Example: Take the values given in example 5. The air at the dew-point contains slightly over 5 grains per cubic foot. At 150° it is capable of containing 73 grains per cubic foot. Consequently, 73-5=68 grains of water which can be evaporated per cubic foot of space at the dew-point when the temperature is raised to 150°. But the latent heat necessary to produce evaporation must be supplied in addition to the heat required to raise the air to 150°.

Example 8. To find the amount of water evaporated during a given change of temperature and humidity:

Example: At 70° suppose the humidity is found to be 64 per cent and at 150° it is found to be 60 per cent. How much water has been evaporated per cubic foot of space? At 70° temperature and 64 per cent humidity there are 5 grains of water present per cubic foot at the dew-point (example 2). At 150° and 60 per cent humidity there are 45 grains present. Therefore, 45-5=40 grains of water which have been evaporated per cubic foot of space, figuring all volumes at the dew-point.

Example 9. To correct readings of the hygrometer for changes in barometric pressure:

A change of pressure affects the reading of the wet bulb. The chart applies at a barometric pressure of 30 inches, and, except for great accuracy, no correction is generally necessary.

Find the relative humidity as usual. Then look for the nearest barometer line (indicated by dashes). At the end of each barometer line will be found a fraction which represents the proportion of the relative humidity already found, which must be added or subtracted for a change in barometric pressure. If the barometer reading is less than 30 inches, add; if greater than 30 inches, subtract. The figures given are for a change of 1 inch; for other changes use proportional amounts. Thus, for a change of 2 inches use twice the indicated ratio; for half an inch use half, and so on.

Example: Dry bulb 67°, wet bulb 51°, barometer 28 inches. The relative humidity is found, by the method given in example 1, to equal 30 per cent. The barometric [241] line — gives a value of 3/100H for each inch of change. Since the barometer is 2 inches below 30, multiply 3/100H by 2, giving 6/100H. The correction will, therefore, be 6/100 of 30, which equals 1.8. Since the barometer is below 30, this is to be added, giving a corrected relative humidity of 31.8 per cent.

This has nothing to do with the vapor pressure (concave) curves, which are independent of barometric pressure, and consequently does not affect the solution of the previous problems.

Example 10. At what temperature must the condenser be maintained to produce a given humidity?

Example: Suppose the temperature in the drying room is to be kept at 150° Fahr., and a humidity of 80 per cent is desired. If the humidity is in excess of 80 per cent the air must be cooled to the dew-point corresponding to this condition (see example 4), which in this case is 141.5°.

Hence, if the condenser cools the air to this dew point the required condition is obtained when the air is again heated to the initial temperature.

Example 11. Determination of relative humidity by the dew-point:

The quantity of moisture present and relative humidity for any given temperature may be determined directly and accurately by finding the dew-point and applying the concave (vapor-pressure) curves. This does away with the necessity for the empirical convex curves and wet-and-dry-bulb readings. To find the dew-point some form of apparatus, consisting essentially of a thin glass vessel containing a thermometer and a volatile liquid, such as ether, may be used. The vessel is gradually cooled through the evaporation of the liquid, accelerated by forcing air through a tube until a haze or dimness, due to condensation from the surrounding air, first appears upon the brighter outer surface of the glass. The temperature at which the haze first appears is the dew-point. Several trials should be made for this temperature determination, using the average temperature at which the haze appears and disappears.

To determine the relative humidity of the surrounding air by means of the dew-point thus determined, find the concave curve intersecting the top horizontal (100 per [242] cent relative humidity) line nearest the dew-point temperature. Follow parallel with this curve till it intersects the vertical line representing the temperature of the surrounding air. The horizontal line passing through this intersection will give the relative humidity.

Example: Temperature of surrounding air is 80; dew-point is 61; relative humidity is 53 per cent.

The dew-point determination is, however, not as convenient to make as the wet-and-dry-bulb hygrometer readings. Therefore, the hygrometer (convex) curves are ordinarily more useful in determining relative humidities.

The Hygrodeik

In Figure 94 will be seen the Hygrodeik. This instrument is used to determine the amount of moisture in the atmosphere. It is a very useful instrument, as the readings may be taken direct with accuracy.

To find the relative humidity in the atmosphere, swing the index hand to the left of the chart, and adjust the sliding pointer to that degree of the wet-bulb thermometer scale at which the mercury stands. Then swing the index hand to the right until the sliding pointer intersects the curved line, which extends downwards to the left from the degree of the dry-bulb thermometer scale, indicated by the top of the mercury column in the dry-bulb tube.

At that intersection, the index hand will point to the relative humidity on scale at bottom of chart (for example see Fig. 94). Should the temperature indicated by the wet-bulb thermometer be 60 degrees, and that of the dry-bulb 70 degrees, the index hand will indicate humidity 55 degrees, when the pointer rests on the intersecting line of 60 degrees and 80 degrees.

The Recording Hygrometer

In Figure 95 is shown the Recording Hygrometer complete with wet and dry bulbs, two connecting tubes and two recording pens and special moistening device for supplying water to the wet bulb.

This equipment is designed particularly for use in connection with dry rooms and dry kilns and is arranged so [243] that the recording instrument and the water supply bottle may be installed outside of the dry kiln or drying room, while the wet and dry bulbs are both installed inside the room or kiln at the point where it is desired to measure the humidity. This instrument records on a weekly chart the humidity for each hour of the day, during the entire week.

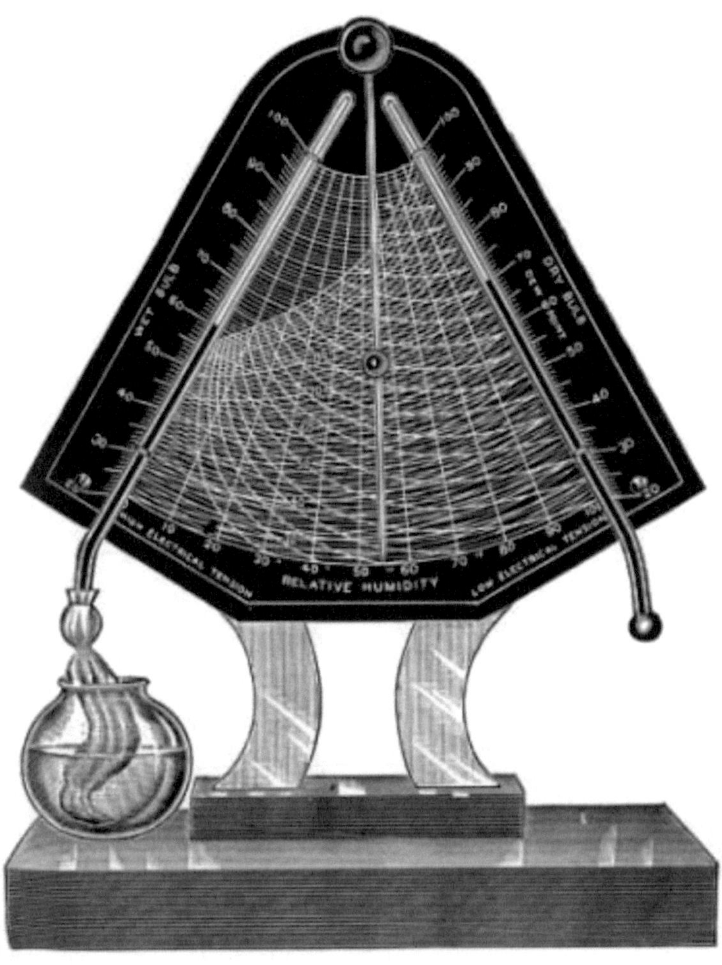

Fig. 94. The Hygrodeik. [244]

The Registering Hygrometer

In Figure 96 is shown the Registering Hygrometer, which consists of two especially constructed thermometers. The special feature of the thermometers permits placing the instrument in the dry kiln without entering the drying room, through a small opening, where it is left for about 20 minutes.

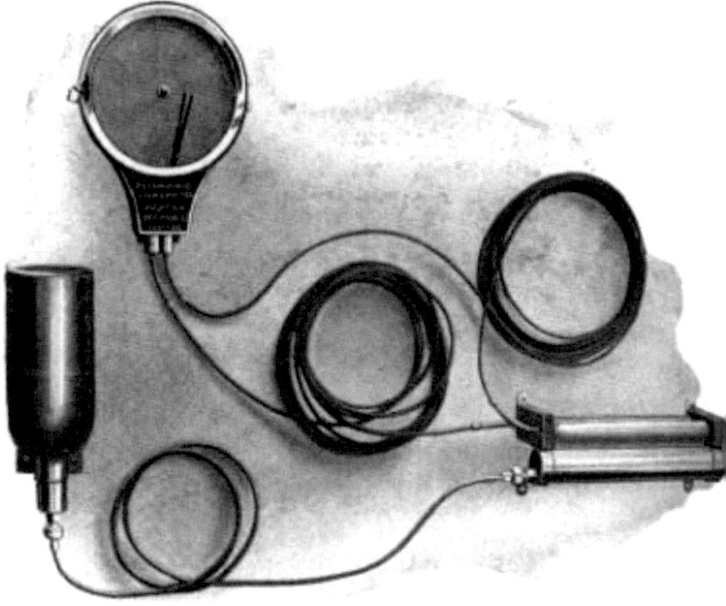

Fig. 95. The Recording Hygrometer, Complete with Wet and Dry Bulbs. This instrument records on a weekly chart the humidity for each hour of the day, during the entire week.

The temperature of both the dry and wet bulbs are automatically recorded, and the outside temperature will not affect the thermometers when removed from the kiln. From these recorded temperatures, as shown when the instrument is removed from the kiln, the humidity can be easily determined from a simple form of chart which is furnished free by the makers with each instrument. [245]

The Recording Thermometer

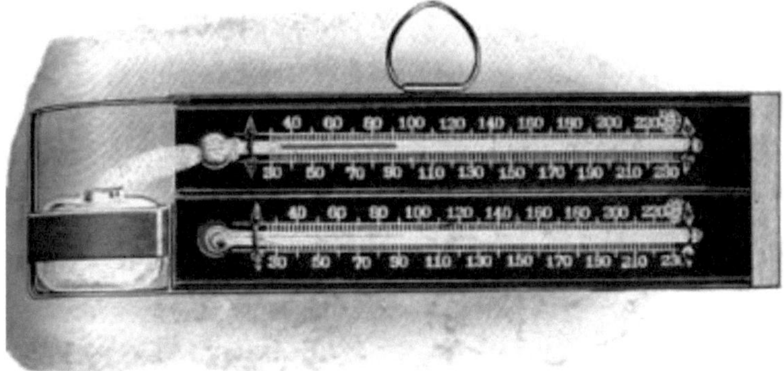

Fig. 96. The Registering Hygrometer.

Fig. 97. The Recording Thermometer.

In Figure 97 is shown the Recording Thermometer for observing and recording the temperatures within a dry kiln, and thus obtaining a check upon its operation. This

[246] instrument is constructed to record automatically, upon a circular chart, the temperatures prevailing within the drying room at all times of the day and night, and serves not only as a means of keeping an accurate record of the operation of the dry kiln, but as a valuable check upon the attendant in charge of the drying process.

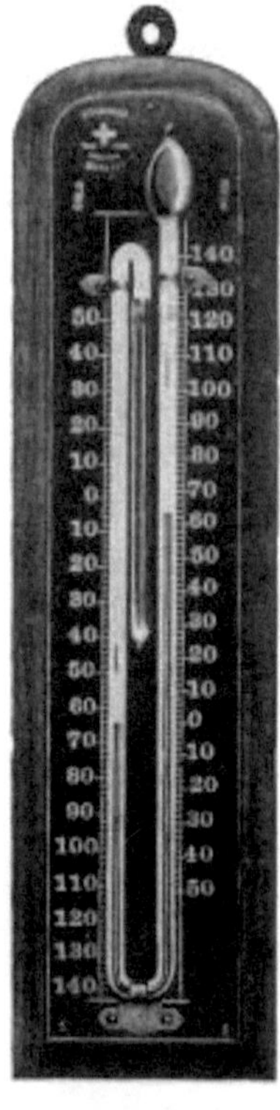

Fig. 98. The Registering Thermometer.

Fig. 99. The Recording Steam-Pressure C

The Registering Thermometer

In Figure 98 is shown the Registering Thermometer, which is a less expensive instrument than that shown in

Figure 97, but by its use the maximum and minimum temperatures in the drying room during a given period can be determined.

The Recording Steam Gauge

In Figure 99 is shown the Recording Steam Pressure Gauge, which is used for accurately recording the steam pressures kept in the boilers. This instrument may be [247] mounted near the boilers, or may be located at any distance necessary, giving a true and accurate record of the fluctuations of the steam pressure that may take place within the boilers, and is a check upon both the day and night boiler firemen.

The Troemroid Scalometer

In Figure 100 is shown the Troemroid Scalometer. This instrument is a special scale of extreme accuracy, fitted with agate bearings with screw adjustment for balancing. The beam is graduated from 0 to 2 ounces, divided into 100 parts, each division representing 1-50th of an ounce; and by using the pointer attached to the beam weight, the 1-100th part of an ounce can be weighed.

Fig. 100. The Troemroid Scalometer.

The percentage table No. II has a range from one half of 1 per cent to 30 per cent and is designed for use where extremely fine results are needed, or where a very small [248] amount of moisture is present. Table No. III ranges from 30 per cent up to 90 per cent. These instruments are in three models as described below.

Model A. (One cylinder) ranges from 1/2 of 1 per cent to 30 per cent and is to be used for testing moisture contents in kiln-dried and air-dried lumber.

Model B. (Two cylinders) ranges from 1/2 of 1 per cent up to 90 per cent and is to be used for testing the moisture contents of kiln-dried, air-dried, and green lumber.

Model C. (One cylinder) ranges from 30 per cent to 90 per cent and is applicable to green lumber only.

Test Samples.—The green boards and all other boards intended for testing should be selected from boards of fair average quality. If air-dried, select one about half way up the height of the pile of lumber. If kiln-dried, two thirds the height of the kiln car. Do not remove the kiln car from the kiln until after the test. Three of four test pieces should be cut from near the middle of the cross-wise section of the board, and 1/8 to 3/16 inch thick. Remove the superfluous sawdust and splinters. When the test pieces are placed on the scale pan, be sure their weight is less than two ounces and more than 13/4 ounces. If necessary, use two or more broken pieces. It is better if the test pieces can be cut off on a fine band saw.

Weighing.—Set the base of the scale on a level surface and accurately balance the scale beam. Put the test pieces on the scale pan and note their weight on the lower edge of the beam. Set the indicator point on the horizontal bar at a number corresponding to this weight, which may be found on the cylinder at the top of the table.

Dry the test pieces on the Electric Heater (Fig. 101) 30 to 40 minutes, or on the engine cylinder two or three hours. Weigh them at once and note the weight. Then turn the cylinder up and at the left of it under the small pointer find the number corresponding to this weight. The percentage of moisture lost is found directly under pointer on the horizontal bar first mentioned. The lower portion on the cylinder Table No. II is an extension of [249] the upper portion, and is manipulated in the same manner except that the bottom line of figures is used for the first weight, and the right side of cylinder for second weight. Turn the cylinder down instead of up when using it.

Examples (Test Pieces)

Model A. Table No. II, Kiln-dried or Air-dried Lumber:

If first weight is 90½ and the second weight is 87, the cylinder table will show the board from which the test pieces were taken had a moisture content of 3.8 per cent.

Model B. Tables No. II and III, Air-dried (also Green and Kiln-dried) Lumber.

If the first weight on lower cylinder is 97 and the second weight is 76, the table will show 21.6 per cent of moisture.

Model C. Table III, Green Lumber:

If the first weight is 94 and the second weight is 51, the table shows 45.8 per cent of moisture.

Keep Records of the Moisture Content

Saw Mills.—Should test and mark each pile of lumber when first piled in the yard, and later when sold it should be again tested and the two records given to the purchaser.

Factories.—Should test and mark the lumber when first received, and if piled in the yard to be kiln-dried later, it should be tested before going into the dry kiln, and again before being removed, and these records placed on file for future reference.

Kiln-dried lumber piled in storage rooms (without any heat) will absorb 7 to 9 per cent of moisture, and even when so stored should be tested for moisture before being manufactured into the finished product.

Never work lumber through the factory that has more than 5 or 6 per cent of moisture or less than 3 per cent.

Dry storage rooms should be provided with heating coils and properly ventilated.

Oak or any other species of wood that shows 25 or 30 per cent of moisture when going into the dry kiln, will take longer to dry than it would if it contained 15 to 20 per cent, therefore the importance of testing before putting into the kiln as well as when taking it out. [250]

The Electric Heater

In Figure 101 is shown the Electric Heater. This heater is especially designed to dry quickly the test pieces for use in connection with the Scalometer (see Fig. 100) without charring them. It may be attached to any electric light socket of 110 volts direct or alternating current. A metal rack is provided to hold the test pieces vertically on edge.

Fig. 101. The Electric Heater.

Turn the test pieces over once or twice while drying.

It will require from 20 minutes to one hour to remove all the moisture from the test pieces when placed on this heater, depending on whether they are cut from green, air-dried, or kiln-dried boards.

Test pieces cut from softwoods will dry quicker than those cut from hardwoods.

When the test pieces fail to show any further loss in weight, they are then free from all moisture content.

BIBLIOGRAPHY [251]

- American Blower Company, Detroit, Mich.
- Imre, James E., "The Kiln-drying of Gum," The United States Dept. of Agriculture, Division of Forestry.
- National Dry Kiln Company, Indianapolis, Ind.
- Prichard, Reuben P., "The Structure of the Common Woods," The United States Dept. of Agriculture, Division of Forestry, Bulletin No. 3.
- Roth, Filibert, "Timber," The United States Dept. of Agriculture, Division of Forestry, Bulletin No. 10.
- Standard Dry Kiln Company, Indianapolis, Ind.
- Sturtevant Company, B. F., Boston, Mass.
- Tieman, H. D., "The Effects of Moisture upon the Strength and Stiffness of Wood," The United States Dept. of Agriculture, Division of Forestry, Bulletin No. 70.
- Tieman, H. D., "Principles of Kiln-drying Lumber," The United States Dept. of Agriculture, Division of Forestry.
- Tieman, H. D., "The Theory of Drying and its Application, etc.," The United States Dept. of Agriculture, Division of Forestry, Bulletin No. 509.
- The United States Dept. of Agriculture, Division of Forestry, "Check List of the Forest Trees of the United States."
- The United States Dept. of Agriculture, Division of Forestry, Bulletin No. 37.
- Von Schrenk, Herman, "Seasoning of Timbers," The United States Dept. of Agriculture, Division of Forestry, Bulletin No. 41.
- Wagner, J. B., "Cooperage," 1910. [252]

GLOSSARY [253]

- **Abnormal.** Differing from the usual structure.
- **Acuminate.** Tapering at the end.
- **Adhesion.** The union of members of different floral whorls.
- **Air-seasoning.** The drying of wood in the open air.
- **Albumen.** A name applied to the food store laid up outside the embryo in many seeds; also nitrogenous organic matter found in plants.
- **Alburnam.** Sapwood.
- **Angiosperms.** Those plants which bear their seeds within a pericarp.
- **Annual rings.** The layers of wood which are added annually to the tree.
- **Apartment kiln.** A drying arrangement of one or more rooms with openings at each end.
- **Arborescent.** A tree in size and habit of growth.
-
- **Baffle plate.** An obstruction to deflect air or other currents.
- **Bastard cut.** Tangential cut. Wood of inferior cut.
- **Berry.** A fruit whose entire pericarp is succulent.
- **Blower kiln.** A drying arrangement in which the air is blown through heating coils into the drying room.
- **Box kiln.** A small square heating room with openings in one end only.
- **Brittleness.** Aptness to break; not tough; fragility.
- **Burrow.** A shelter; insect's hole in the wood.
-
- **Calorie.** Unit of heat; amount of heat which raises the temperature.
- **Calyx.** The outer whorl of floral envelopes.
- **Capillary.** A tube or vessel extremely fine or minute.

- **Case-harden.** A condition in which the pores of the wood are closed and the outer surface dry, while the inner portion is still wet or unseasoned.
- **Cavity.** A hollow place; a hollow.
- **Cell.** One of the minute, elementary structures comprising the greater part of plant tissue.
- **Cellulose.** A primary cell-wall substance. [254]
- **Checks.** The small chinks or cracks caused by the rupture of the wood fibres.
- **Cleft.** Opening made by splitting; divided.
- **Coarse-grained.** Wood is coarse-grained when the annual rings are wide or far apart.
- **Cohesion.** The union of members of the same floral whorl.
- **Contorted.** Twisted together.
- **Corolla.** The inner whorl of floral envelopes.
- **Cotyledon.** One of the parts of the embryo performing in part the function of a leaf, but usually serving as a storehouse of food for the developing plant.
- **Crossers.** Narrow wooden strips used to separate the material on kiln cars.
- **Cross-grained.** Wood is cross-grained when its fibres are spiral or twisted.
-
- **Dapple.** An exaggerated form of mottle.
- **Deciduous.** Not persistent; applied to leaves that fall in autumn and to calyx and corolla when they fall off before the fruit develops.
- **Definite.** Limited or defined.
- **Dew-point.** The point at which water is deposited from moisture-laden air.
- **Dicotyledon.** A plant whose embryo has two opposite cotyledons.
- **Diffuse.** Widely spreading.
- **Disk.** A circular, flat, thin piece or section of the tree.

- **Duramen.** Heartwood.
-
- **Embryo.** Applied in botany to the tiny plant within the seed.
- **Enchinate.** Beset with prickles.
- **Expansion.** An enlargement across the grain or lengthwise of the wood.
-
- **Fibres.** The thread-like portion of the tissue of wood.
- **Fibre-saturation point.** The amount of moisture wood will imbibe, usually 25 to 30 per cent of its dry-wood weight.
- **Figure.** The broad and deep medullary rays as in oak showing when the timber is cut into boards.
- **Filament.** The stalk which supports the anther.
- **Fine-grained.** Wood is fine-grained when the annual rings are close together or narrow.
-
- **Germination.** The sprouting of a seed.
- **Girdling.** To make a groove around and through the bark of a tree, thus killing it. [255]
- **Glands.** A secreting surface or structure; a protuberance having the appearance of such an organ.
- **Glaucous.** Covered or whitened with a bloom.
- **Grain.** Direction or arrangement of the fibres in wood.
- **Grubs.** The larvae of wood-destroying insects.
- **Gymnosperms.** Plants bearing naked seeds; without an ovary.
-
- **Habitat.** The geographical range of a plant.
- **Heartwood.** The central portion of tree.
- **Hollow-horning.** Internal checking.
- **Honeycombing.** Internal checking.

- **Hot-blast kiln.** A drying arrangement in which the air is blown through heating coils into the drying room.
- **Humidity.** Damp, moist.
- **Hygroscopicity.** The property of readily imbibing moisture from the atmosphere.
-
- **Indefinite.** Applied to petals or other organs when too numerous to be conveniently counted.
- **Indigenous.** Native to the country.
- **Involute.** A form of vernation in which the leaf is rolled inward from its edges.
-
- **Kiln-drying.** Drying or seasoning of wood by artificial heat in an inclosed room.
-
- **Leaflet.** A single division of a compound leaf.
- **Limb.** The spreading portion of the tree.
- **Lumen.** Internal space in the spring- and summer-wood fibres.
-
- **Median.** Situated in the middle.
- **Medulla.** The pith.
- **Medullary rays.** Rays of fundamental tissue which connect the pith with the bark.
- **Membranous.** Thin and rather soft, more or less translucent.
- **Midrib.** The central or main rib of a leaf.
- **Moist-air kiln.** A drying arrangement in which the heat is taken from radiating coils located inside the drying room.
- **Mottle.** Figure transverse of the fibres, probably caused by the action of wind upon the tree.
-
- **Non-porous.** Without pores.
-

- **Oblong.** Considerably longer than broad, with flowing outline.
- **Obtuse.** Blunt, rounded. [256]
- **Oval.** Broadly elliptical.
- **Ovary.** The part of the pistil that contains the ovules.
-
- **Parted.** Cleft nearly, but not quite to the base or midrib.
- **Parenchyma.** Short cells constituting the pith and pulp of the tree.
- **Pericarp.** The walls of the ripened ovary, the part of the fruit that encloses the seeds.
- **Permeable.** Capable of being penetrated.
- **Petal.** One of the leaves of the corolla.
- **Pinholes.** Small holes in the wood caused by worms or insects.
- **Pistil.** The modified leaf or leaves which bear the ovules; usually consisting of ovary, style and stigma.
- **Plastic.** Elastic, easily bent.
- **Pocket kilns.** Small drying rooms with openings on one end only and in which the material to be dried is piled directly on the floor.
- **Pollen.** The fertilizing powder produced by the anther.
- **Pores.** Minute orifices in wood.
- **Porous.** Containing pores.
- **Preliminary steaming.** Subjecting wood to a steaming process before drying or seasoning.
- **Progressive kiln.** A drying arrangement with openings at both ends, and in which the material enters at one end and is discharged at the other.
-
- **Rick.** A pile or stack of lumber.
- **Rift.** To split; cleft.

- **Ring shake.** A large check or crack in the wood following an annual ring.
- **Roe.** A peculiar figure caused by the contortion of the woody fibres, and takes a wavy line parallel to them.
-
- **Sapwood.** The outer portions of the tree next to the bark; alburnam.
- **Saturate.** To cause to become completely penetrated or soaked.
- **Season checks.** Small openings in the ends of the wood caused by the process of drying.
- **Seasoning.** The process by which wood is dried or seasoned.
- **Seedholes.** Minute holes in wood caused by wood-destroying worms or insects.
- **Shake.** A large check or crack in wood caused by the action of the wind on the tree.
- **Shrinkage.** A lessening or contraction of the wood substance. [257]
- **Skidways.** Material set on an incline for transporting lumber or logs.
- **Species.** In science, a group of existing things, associated according to properties.
- **Spermatophyta.** Seed-bearing plants.
- **Spring-wood.** Wood that is formed in the spring of the year.
- **Stamen.** The pollen-bearing organ of the flower, usually consisting of filament and anther.
- **Stigma.** That part of the pistil which receives the pollen.
- **Style.** That part of the pistil which connects the ovary with the stigma.
-
- **Taproot.** The main root or downward continuation of the plant axis.

- **Temporary checks.** Checks or cracks that subsequently close.
- **Tissue.** One of the elementary fibres composing wood.
- **Thunder shake.** A rupture of the fibres of the tree across the grain, which in some woods does not always break them.
- **Tornado shake.** (See Thunder shake.)
- **Tracheids.** The tissues of the tree which consist of vertical cells or vessels closed at one end.
-
- **Warping.** Turning or twisting out of shape.
- **Wind shake.** (See Thunder shake.)
- **Working.** The shrinking and swelling occasioned in wood.
- **Wormholes.** Small holes in wood caused by wood-destroying worms.
-
- **Vernation.** The arrangement of the leaves in the bud.
- **Whorl.** An arrangement of organs in a circle about a central axis. [258]